CONTEMPORARY RURAL SYSTEMS IN TRANSITION

Volume 1

Agriculture and Environment

CONTEMPORARY RURAL SYSTEMS IN TRANSITION

VOLUME 1

Agriculture and Environment

Edited by

I.R. Bowler, C.R. Bryant and M.D. Nellis

C·A·B International

C·A·B International
Wallingford
Oxon OX10 8DE
UK

Tel: Wallingford (0491) 32111
Telex: 847964 (COMAGG G)
Telecom Gold/Dialcom: 84: CAU001
Fax: (0491) 33508

A catalogue record for this book is available from the British Library

ISBN 0 85198 811 3 (Volume 1)
ISBN 0 85198 812 1 (Volume 2)

ISBN 0 85198 813 X (Two-volume set)

Typeset by Intype, London
Printed and bound in the UK by Redwood Press Ltd., Melksham

Contents

Notes on Contributors

Ian Bowler is a Senior Lecturer in the Department of Geography at the University of Leicester, having previously worked at the University of Strathclyde. He holds BA and PhD degrees from the University of Liverpool and has a specialist research interest in state intervention in agriculture. His books include *Agriculture under the Common Agricultural Policy* (1986, Manchester University Press, Manchester), and *The Geography of Agriculture in Developed Market Economies* (Editor) (1993, Longman, London). He is a former Chair of the Rural Geography Study Group of the Institute of British Geographers and a member of the IGU Commission on 'Changing Rural Systems'.

Christopher Bryant has been a full professor in the Département de Géographie at the Université de Montréal since 1990, after being with the University of Waterloo in southern Ontario since 1970; he was also Director of the Economic Development Program at Waterloo between 1984 and 1990. He holds a PhD from the London School of Economics and Political Science. His primary research interests are in urban fringe agriculture and local and community economic development. He has an active research programme, with field components in France, Québec and Ontario. Together with L.H. Russwurm and A.G. McLellan, he published *The City's Countryside* (1982, Longman, London), and *Agriculture in the City's Countryside* (1992) co-authored with T.R.R. Johnston (Belhaven Press, London). He has published several dozen articles and chapters on urban fringe agriculture and community development. His research programme has been supported principally by the Social Sciences and Humanities Research Council of Canada.

Quentin Chiotti recently completed his PhD at the University of Western Ontario in London, Ontario. In his dissertation, he analysed the differential development between the evolving policy and institutional arrangements for the dairy and beef livestock sectors in Canada. Most of the research for the chapter in this volume was conducted during a two-year sessional position at the University of Guelph, Ontario. He is currently a SSHRC Postdoctoral Fellow at Carleton University, Ottawa, investigating the impact of the Canada–United States Free Trade Agreement and the General Agreement on Tariffs and Trade (GATT) negotiations on the dairy, poultry and wine agri-food sectors in Canada.

Richard Cobb is a graduate of the School of Environmental Sciences at the University of East Anglia (Norwich). He is currently a postgraduate in the School but previously worked on the research project reported in Chapter 11.

Janel Curry-Roper received her PhD in geography from the University of Minnesota, and is presently Associate Professor of geography at Central College, Pella, Iowa. She has published in a wide variety of geography and natural resource journals; her research interests are in rural geography, natural resources and agricultural policy. She has also carried out research into belief systems and their relationship to environmental ethics.

Pierre Deslauriers is a doctoral candidate in the Département de Géographie at the Université de Montréal. He is currently writing a thesis on structural change in agriculture and farmers' adaptive behaviour in Montréal's urban fringe. He holds a Masters degree in Geography from the Université de Montréal. His main research interest is in urban fringe agriculture, but his interests also include the broader process of demographic and land use change around cities and the contemporary restructuring processes in agriculture. He has held fellowships from the Canada Mortgage and Housing Corporation and the Social Sciences and Humanities Research Council of Canada. Since 1988, he has taught courses both at the Université de Montréal and Concordia University in economic geography.

Andrew Gilg is a Senior Lecturer in Geography at the University of Exeter with research interests in rural planning, alpine tourism, and land use planning in Switzerland and North America. He is editor of *Progress in Rural Policy and Planning*, the successor to the *Countryside Planning Yearbook* (1980–1986) and the *International Yearbook of Rural Planning* (1987–1988) which he also edited. His other publications include: *An Introduction to Rural Geography* (1985, Edward Arnold, London); (with M. Blacksell) *The Countryside: Planning and Change* (1981, Allen and Unwin, London); *Countryside Planning* (1979, Methuen, London);

Countryside Planning Policies for the 1990s (1991, CAB, Wallingford); and editor of *Restructuring the Countryside: Environmental Policy in Practice* (1992, Ashgate, Aldershot).

Lisa Harrington is an Assistant Professor of geography in the Geology/Geography Department at Eastern Illinois University. Her research focuses on places far from cities, especially as regards reserved lands. The establishment and use of National Parks and Wilderness areas has been a particular interest, together with topics in natural resources and biogeography.

Stewart Hilts is Associate Professor, with a cross-appointment with the Department of Land Resource Science and the University School of Rural Planning and Development, at the University of Guelph, in Ontario, Canada. He is currently serving as Director of the Centre for Land and Water Stewardship at the University. Professor Hilts' research interests focus on the relationship between the conservation practices of rural landowners and government policies on the environment and agriculture. He is particularly interested in the development of policies or programmes that can enhance private land stewardship. Previously he has published three books: *Islands of Green – Natural Heritage Protection in Ontario; The Guelph Seminars on Sustainable Development*; and *Natural Heritage Landowner Contact Training Manual*; as well as several chapters in other books, and numerous technical reports.

Brian Ilbery, Reader in Geography at Coventry University, has research interests in agricultural change and policy, farm diversification and alternative farming systems. His publications include: *Agricultural Geography – A Social and Economic Analysis* (1985, Oxford University Press, Oxford); *Agricultural Change – France and the EEC* (1988, John Murray, London); *Location and Change – Perspectives on Economic Geography* (1990, Oxford University Press, Oxford); and *Agricultural Change in Great Britain* (1991, Oxford University Press, Oxford). A large number of original research papers have also been published in geographical and other scientific journals.

Claude Marois obtained his PhD in geography at Laval University in 1980, and joined the Département de Géographie at the Université de Montréal in 1977. He has been Associate Professor in that department since 1986. A member of several geography and regional science associations, he was appointed as a member of the IGU Commission on Population Geography. He teaches population and social geography as well as statistical applications and research methodology. His research interests include the impacts of urban sprawl on the urban fringe and the adaptation of communities in those areas. He is also interested in the geography of ethnic residential segregation and social structures and changes in land use. He has published some 30 articles in scholarly journals and as many chapters in books, monographs and reports. He has also given scholarly presentations at conferences in Canada, France and the United States, and has acted as consultant to various public agencies.

Terry Marsden is Lecturer in Geography in the School of Geography, University of Hull. He is editor of the *Critical Perspectives on Rural Change* series (David Fulton Ltd, London) and has research interests in rural land development, agricultural change, fresh food systems and family farming.

Richard Munton is Professor and Head of Department of Geography at University College London. He is Chair of the NERC Land Use Research Coordinating Committee and has research interests in land development processes, property rights and family farming. His books include *London's Green Belt – Containment in Practice* (1983, George Allen and Unwin, London).

Darrell Napton was born and raised on a farm in the US Midwest, where his family continues to farm and that has been in the family for 12 decades. His life-long interest in farming and rural land use was reflected in his PhD research at the University of Minnesota, and more recently in his teaching and research as Associate Professor at South Dakota State University (Brookings). His recent research has focused on farmland protection policies and agricultural diversification.

Duane Nellis is currently Professor and Head of Geography, Presidential Lecturer, and Director of the Institute for Social and Behavioral Research at Kansas State University. Born in Spokane, Washington, Nellis received his BSc degree in geography from Montana State University in 1976, MSc and PhD from Oregon State University in 1977 and 1980 respectively. He has authored numerous articles on United States and African rural resource systems, and the application of remote sensing and geographic information systems to rural resource analysis. He has received numerous awards, including the Institute of British Geographers Young Research Worker Award, and the Association of American Geographers (AAG) Contemporary Agriculture and Rural Land Use John Fraser Hart Award for Research Excellence. He is also past Chair of the Contemporary Agriculture and Rural Land Use Specialty Group of the AAG.

John Pierce is Chair in the Department of Geography at Simon Fraser University, Canada. He is author (with O.J. Furuseth) of *Agricultural Land in an Urban Society* (1982, AAG Research Publications in Geography, Washington DC), and *The Food Resource* (1990, Longman, London). He is a former Chair of the Rural and Urban Fringe Study Group of the Canadian Association of Geographers, and has research interests in food production, the environmental implications of agricultural policy and farmland preservation strategies.

Timothy Rickard was born in Swansea, Wales, in 1942. He received his BA degree from the University College of Wales, Aberystwyth, his MA from the University of Nebraska and his PhD from the University of Kansas – all in geography. Author of numerous articles on agricultural geography, he is Chair of the Contemporary Agriculture and Rural Land Use Specialty Group of the Association of American Geographers. He received the New England St Lawrence Valley Geographical Society Distinguished Service Award in 1986, for research on agricultural preservation and long-term editorial duties, including a book on geography in the region. Chair of the Geography Department at Central Connecticut State University for 11 years and Professor of Geography, he currently serves as Assistant for International Education to the Academic Vice President.

Rebecca Roberts holds a PhD from Oregon State University and is Assistant Professor of Geography at The University of Iowa. Her research interests centre on the roles played by the processes and relationships of production in the development of environmental conflicts, specifically the intersection between agricultural production and water resource problems. Her recent research has focused on groundwater depletion by irrigation on the High Plains of Texas and New Mexico, and on groundwater contamination by agricultural chemicals in the Midwest. Her work has appeared in *Economic Geography, Political Geography Quarterly, Journal of Rural Studies, Environment and Planning A, Water Resources Bulletin* and other journals.

John Tarrant holds BSc and PhD degrees from the University of Hull. His first academic post was in the Department of Geography at University College Dublin and he joined the School of Environmental Sciences at the University of East Anglia when the School was founded in 1968. Following promotions to Senior Lecturer and Reader, he is currently Deputy Vice-Chancellor of the University. Research has been undertaken in many developing and developed countries, and he has held posts of Research Fellow at the International Food Policy Research Institute (Washington, DC) and at the Food Research Institute in Stanford University and has been Harris Visiting Professor in the College of Geosciences at the Texas A and M University. His many research publications include *Food Policy* (1980, Wiley, Chichester).

Michael Troughton received a BLitt from Oxford and his PhD from the University of Western Ontario. He is currently Full Professor in the Department of Geography at the University of Western Ontario. His research and teaching interests are in agricultural and rural geography, including rural resource management. His publications have dealt with Canadian agriculture, the rural-urban fringe and marginal areas, agricultural industrialization and institutional arrangements. He is currently Chair of the IGU Commission on 'Changing Rural Systems' and President of the Canadian Association of Geographers.

Iain Wallace studied at Oxford and Bristol before taking up a position at Carleton University, Ottawa, where he is a Professor of Geography. His research interests lie in the fields of: global economic change; resource-based industries, particularly agribusiness; the relationship between the economy and the environment; and the economic geography of Canada. He is the author of *The Global Economic System* (1990, Routledge, London).

Neil Ward is a Research Officer in the Department of Geography at University College, London. His research interests include agricultural adjustment, family farming, pollution and water quality.

Charles Watkins is Lecturer in Geography at the University of Nottingham and was formerly Assistant Director of the Centre for Rural Studies at the Royal Agricultural College, Cirencester. His main research interests include rural geography, land management and landscape history. He is currently researching the legal and clerical professions in rural Britain, the management of woodland and grassland and game management. Recent books include *Woodland Management and Conservation* (1990, David and Charles, Newton Abbot). He is co-author of *Justice Outside the City* (1991, Longman, London) and *Church and Religion in Rural England* (1991, Clark, London), co-editor of *People in the Countryside: Studies of Social Change in Rural Britain* (1991, Paul Chapman, London), and editor of *Ecological Effects of Afforestation* (1992, CAB International, Wallingford).

Preface

In August 1991, an international conference was convened in the United Kingdom by Dr Ian Bowler: the conference brought together rural geographers from Canada, the United Kingdom and the United States. The focus of the conference was the rapid and significant transition of rural areas in developed countries over the previous decade, a transition sometimes referred to as the 'restructuring' of rural economy and society. The conference reviewed a wide range of issues as regards contemporary changes in rural areas, and adopted the descriptive term 'rural system' to encompass and integrate the totality of rural economy and society.

The two volumes of this book are based on the papers presented at the conference. Volume 1 contains chapters on agriculture and the environment: (i) developing organizational structures in the food supply system; (ii) changing farm business structures; (iii) the environmental impact of modern agriculture; (iv) emerging agricultural policy issues; and (v) the growing concern with 'sustainable agriculture'. In Volume 2, the chapters consider aspects of the wider rural economy and society: (i) the emergence of new socioeconomic issues; (ii) changes in the structure of rural society; (iii) trends in countryside recreation and tourism; (iv) the changing employment structure; and (v) development strategies for rural communities.

Each chapter in the book discusses broad processes and structural changes that are common to all rural systems in developed countries; however, different geographical contexts are used to illustrate the uneven development of those systems under the contemporary transition of rural areas. Contributors have drawn upon their varying research experiences in Canada, the UK and the USA: they review the relevant literature for their national contexts and illustrate their arguments from original research. The resulting book, in its two volumes, covers a wide range of issues in the contemporary transition of rural systems and provides a contrasting set of international perspectives.

Acknowledgement

The editors wish to thank the Economic and Social Research Council (UK) for financial support of the conference, the papers from which form the basis of this book.

ABBREVIATIONS

AAG	Association of American Geographers
ALURE	Alternative Land Uses and the Rural Economy
BSU	Business Size Unit
CAP	Common Agricultural Policy
CCA	Canadian Cattlemen's Association
CCC	Commodity Credit Corporation
CCREM	Canadian Council of Resource and Environment Ministers
CDC	Canadian Dairy Commission
CIT	Canadian Import Tribunal
CRP	Conservation Reserve Program
C$	Canadian dollar
CUSTA	Canada–United States Trade Agreement
CWB	Canadian Wheat Board
DOE	Department of Energy (USA)
EC	European Community
EPA	Environment Protection Agency
ESA	Environmentally Sensitive Area
FACTA	Food, Agriculture, Conservation and Trade Act
FDGS	Farm Diversification Grant Scheme
FWS	Farm Woodland Scheme
GATT	General Agreement on Tariffs and Trade
IGU	International Geographical Union
JAPA	Journal of the American Planning Association
LESA	Land Evaluation and Site Assessment
LFA	Less Favoured Area
LTER	Long-Term Ecological Reserves
MAFF	Ministry of Agriculture Fisheries and Food
M&S	Marks and Spencer
NCC	Nature Conservancy Council
NIMBY	Not in my back-yard

OECD	Organization for Economic Cooperation and Development
OMAF	Ontario Ministry of Agriculture and Food
OMMB	Ontario Milk Marketing Board
PEI	Prince Edward Island
PDR	Purchase of Development Rights
RNA	Research Natural Areas
SCS	Soil Conservation Service
SSSI	Site of Special Scientific Interest
TDR	Transfer of Development Rights
TPO	Tree Preservation Order
UK	United Kingdom
US(A)	United States (of America)
USDA	United States Department of Agriculture
US$	United States dollar
USSR	Union of Soviet Socialist Republics (now the Commonwealth of Independent States)

INTRODUCTION

Rural areas have been undergoing rapid change as their economies and societies restructure, partially at least in response to changes in the broader socioeconomic system. Structures and internal and external relationships have been modified in relation to powerful forces that transcend all scales of analysis. In Volume One, the accent is placed on changing agricultural structures and relationships, and a significant thread deals with the links between agriculture and environment. In Volume Two, the emphasis shifts from farming, which remains the major activity and land use in many rural areas, to a consideration of certain aspects of the broader rural economy and society.

Two recurrent themes run through most discussions of rural change: uneven development and the consequences of and responses to uneven development. Uneven development, in terms of differing levels of socioeconomic health, arises from processes of capital accumulation and geographic and non-geographic concentrations of power and wealth, all in the context of limited financial resources. Uneven development is fuelled by flows of capital and labour, and is reinforced by uneven patterns of competitiveness and flows of goods and services. It is manifested in geographic and non-geographic concentrations of wealth and power, and their opposites, poverty, disenfranchisement and lack of control. Patterns and processes of uneven development have always existed, but in the last quarter of the 20th century a critical question is posed for Western countries: that is whether or not the patterns of uneven development are strengthening and becoming more accentuated. In the context of rural areas, which have long been associated with economic stagnation in the more peripheral parts of the economic system, are these areas becoming increasingly vulnerable? Are their economic and social structures becoming increasingly marginalized? Furthermore, what sorts of responses and strategies have been followed to try and modify the internal and external relationships of rural areas? And what sorts of impacts have they had?

Although the chapters in the two volumes of this book do not address the whole range of issues, they touch on most of them. This introductory chapter develops a framework with which to view the contemporary shifts in the rural economies and societies of the Western World and which serves as a backdrop to the contributions in these volumes.

The dynamics of rural change and uneven development are considered first, emphasizing the ties between rural areas and the other parts of our social and economic systems. This theme is also reiterated in a discussion of the heterogeneous nature of rural areas, and the different forces of change that have influenced their transformation. With this as a backdrop, consideration is given to the results of these transformations and some selected responses to them, drawing upon the various contributions to Volume One.

The Dynamics of Rural Change and Uneven Development

Underlying uneven development processes are changes in activities – economic, social and political – that result from modification of socioeconomic systems of production and of the systems of exchange within which all activities function. Different socioeconomic systems of production are characterized by different technical, social, financial, economic and behavioural structures and patterns of interaction. Different systems dominate at different times: for example, at the coarsest level of classification, we can identify subsistence production, artisanal production, capitalistic production and more collectivist modes of organization of production (see Malassis, 1958).

Systems change and, over longer periods of time, are replaced. When the seeds of a newly emerging system are concentrated geographically, the scene is ripe for differential development of the dominant system over geographic space and through time. Hence, at least one of the types of areas where we might expect to see significant developments in capitalistic forms of rural activities is in close proximity to the urban–industrial complex.

Socioeconomic systems of production function through various systems of exchange or interaction (for example, market systems for produce, capital or labour), and these systems can function at a variety of different scales. As the macrosystems change, so do the systems of exchange that animate them, and vice versa. Resulting from this are changing relationships between rural areas (and their activities) and other parts of the economic and social system.

The Heterogeneity of Rural Areas

Many discussions of the nature of change in rural economies and societies oversimplify the situation by talking of 'rural' areas as if they were stand-alone entities and as if they were homogeneous. In relation to these two points, the following should be noted:

1. A 'rural' area is not necessarily a functioning socioeconomic or political unit. At the very least, it is important to appreciate the links that exist between many rural areas and the small towns embedded in them, frequently serving as service centres for rural market areas.
2. Furthermore, 'rural' areas can be differentiated on the basis of a number of dimensions, which are essential for understanding the results of and responses to the transformation of rural areas. Some of the more important differentiating characteristics of rural areas are:
(i) the degree of integration of rural areas with the urban–industrial complex;
(ii) the nature of the economic base of the area (for example, the natural resource base, tourism, services and manufacturing);
(iii) the social and demographic structure of the area and the community or communities present;
(iv) the nature of the political organization of the area and the aspirations and capacities of the communities in relation to their abilities for self-governance and management.

The principal reason for emphasizing the heterogeneity of rural areas is because such differences can be critical in understanding patterns of responses from rural areas and the appropriate strategies for dealing with 'rural' issues. Furthermore, not only is there heterogeneity among rural areas, there are also significant differences between groups of 'rural' people even within the same geographic community or locality. Once again, these differences are important in understanding issues, patterns of responses and appropriate strategies, both at the individual or micro level and at the community and state level.

The Forces of Change

Fundamental to the notion of systems of exchange introduced above is the idea that the interactions reflect the supply and demand for different types of interaction. In the context of economic systems of exchange, different types of interaction reflect the values placed on the goods and services involved in those interactions; the functioning of economic systems of exchange is invariably associated with different socioeconomic modes of production and can also have impacts on social and political systems of exchange. For instance, the operating of the product, labour and capital markets leads to uneven development geographically, giving rise to changes in social systems of exchange within rural communities – for example, where there is out-migration and the development of pockets of poverty.

At the most general level, interactions can be modified by changes in values, institutions and technologies. Changes in values undoubtedly constitute the most fundamental set of changes, being tied to changing demographic structures and migration patterns, as well as underlying the other two categories – institutional and technological changes. During the 1980s, it became fashionable to talk of the broad forces of change in society as 'mega-trends' (see Naisbitt, 1982), particularly in relation to the transformation of society from an industrial to an information and services, or postindustrial, society.

It is easy to think of these forces or trends as somehow removed from the individual, the firm or farm and the rural community, and therefore to think of these as requiring some adjustment, reaction or adaptation on the part of the individuals, firms, farms and communities. However, and this is important from the rural area or community perspective, it should not be forgotten that such trends are composed of the decisions of many individuals and firms. Influences can be transmitted upwards through the systems of exchange as well as downwards, thus emphasizing the opportunity for some proactive behaviour on the part of the elemental decision-making units in the socioeconomic system. Of course, the opportunities for engaging in proactive behaviour vary between rural areas or communities and between individuals in those areas.

What have been the main forces of change leading to transitions in the social and economic system and how have these been translated into the restructuring of rural systems? It is useful to think of the various forces or trends associated with the development of postindustrial society as falling into three categories: (i) the emergence of 'new needs' in society; (ii) the development of new transportational and communications technology; and (iii) the development of new technology affecting production processes (Bryant, 1989). All of these forces have influenced the restructuring of rural systems, as well as many other facets of the organization of society and economy.

'New needs' are linked to a variety of value changes that have occurred and continue to occur in society. As disposable income has increased and the population has, on average, become more 'sophisticated', demands have increased for a greater range of services as well as for goods with a considerable value-added component to them. Values regarding life-styles have changed with, for instance, greater values being placed upon the quality of the rural environment, outdoor recreation and living in a rural community. One outcome has been a resurgence of interest in many countries in the 1980s in the potential for tourist development to bring greater income security to many rural communities. Other value changes include the greater attention being paid to personal health, to the quality of food products and to the environment generally. All of these, potentially, have impacts on rural socioeconomic systems. Some of these changes have the potential to contribute to the rejuvenation of the economic systems of some rural areas, whereas in others they can create difficulties. One need only think of the developing market for paper produced by processes that are chlorine-free to realize the far-reaching impact that this will have on the paper industry and the resource hinterlands involved, depending upon the ability of the companies to modify or change their production technology.

It is also possible to include under new needs the pleas for deregulation and less direct government intervention in many countries. In North America, the Canada–US Free Trade Agreement (CUSTA) can perhaps be placed in this context, and it certainly has far-reaching implications for many activities in rural areas. Similarly, the latest round of negotiations concerning GATT hold significant implications for agricultural activities in many countries.

Changes in transportation and communications technology have been central to the development of postindustrial society. An important part of the changes in transportation technology is, of course, intimately associated with the rise of industrial society. These changes that permitted the bulk transport of goods and produce over long distances already lie at the root of many of the changes in rural employment over the course of the 20th century. Impacts have included the expansion of the geographic limits of production with which any particular economic activity has had to contend, increased competition and greater regional specialization in some activities. More recent changes in transportation technology, including containerization, have simply reinforced many of the patterns of change already underway. In some cases the result has been to strengthen particular regional economies, in others it has helped to undermine them, particularly when they have been associated with 'consolidation and rationalization' processes in the shipping of agricultural produce. Transportation changes have, therefore, altered the boundaries of the systems of exchange that many rural activities function within.

The surge in developments in communications technology is perhaps the epitome of postindustrial society. The 'fax', telecommunications and extensive computer-based networks for the transfer of information have all contributed to the development of the 'global village' and the globalization generally of many aspects of the economic system – for instance, the globalization of capital. These changes are all information transfer or processing technologies. They typify the importance of knowledge as capital in today's society. On the one hand, they permit the decentralization of certain types of activities over huge distances; however, decentralization of the service activities that are dependent upon such technologies has not occurred to the extent that is physically possible. They have, however, had a significant impact on the evolving settlement structures and evolving employment patterns around many major metropolitan areas. The potential impact is even greater: telecommuting centres permit people to live in a decentralized settlement system and to go to 'work' several days a week in a local centre; the centre houses communication equipment and information processing equipment that a variety of companies and agencies might use or own collectively, thus eliminating the need to commute daily to a centralized office.

Many of the changes in production technology can also be traced back to the rise of industrial society. Most notable here was the development of massive scale economies in the production process of many sectors and the substitution of capital for labour. This form of technological change, together with the transportation developments that often went hand in hand with them, has underlain the massive changes in business structure in primary industries. Increases in 'optimal' business size in the face of relatively stagnant markets, together with the rise of competitors in other parts of the world, have led to plant closures and 'downsizing' of the labour force. Furthermore, the fact that, excepting for agriculture, many of the primary activities are internationally owned means that performance measurements in relation to capital invested in such enterprises are determined on an international basis, not a regional or even a national one. In agriculture, the changes have been no less dramatic, having led to widespread rural depopulation for many agricultural areas in North America and Western Europe.

On the other hand, while these same labour-saving technologies continue to play a significant role in some sectors, technological change in other sectors has permitted the growth and development of smaller and medium-sized enterprises. Examples include some sectors of the food-processing industry, where a combination of technology that favours (or at least does not penalize) small-to-medium-size business and a more differentiated market for food products, has encouraged diversification in production. Other examples include the development of hundreds of small desk-top publishing enterprises and other service-related activities where there is no need for the entrepreneur to come into face-to-face contact with the customer. Although still largely concentrated in the major metropolitan areas, there are opportunities in these tendencies for some types of rural areas to derive benefits.

In areas where the rural economy is dominated by resource activities such as forestry and mining, rural economies have been subjected to downward pressures consequent upon technological change both in terms of production processes and in transportation, the accompanying globalization of markets and capital circuits, and corporate restructuring in relation to these. Agriculture has fared little better; this production sector is still highly fragmented but the nature of the markets faced by many farm areas has seemed to remove them from any control over their destinies.

These macro-patterns and forces form the backdrop against which changes in rural economies and societies beyond metropolitan spheres of influence must be seen. They are also present in rural areas within the metropolitan regions; traditional rural activities there, especially farming, still represent important contributors to national income, even though absolute employment in these activities has decreased and other forms of employment have generally replaced this decline.

A Summary of the Chapters in Volume One

The transition of rural systems can be examined in a variety of ways. Here attention is confined to agriculture and the rural environment and to a number of themes. On the last aspect, the themes cover the evolving agri-food system, diversification of the farm business, the changing relationship between agriculture and the environment, the development of agricultural policy and the emergence of 'sustainable' agriculture as a policy option.

The evolving agri-food system

An increasing proportion of the agri-food chain is only indirectly linked to agricultural production, yet most developments for food and beverage markets have some implications for farm-based production. The result has been a major transition in the spatial reorganization of rural systems. In countries such as the United Kingdom and the United States there has been an acceleration of corporate concentration, changing consumer demands and more integrated continental markets. In Chapter 1, Wallace addresses selective features of this process of international or global (see p. 5) restructuring. In Canada, for example, agriculture is in the midst of a fundamental restructuring that is part of a broader shift to an industrial model. In recent years, the agricultural system has been dominated by agribusiness and government, with significant implications relative to the stability of the agricultural environment. Governments have facilitated the integration of the system to the point where farmers have become dependent on subsidies to remain in business. The result, as summarized by Troughton in Chapter 2, is a degrading of the rural physical environment, and a decoupling of farms and farm populations from the rural community.

The implementation of free-trade agreements (see p. 5), in an attempt at more integrative continental markets, has also had a profound impact on the rural systems of Canada, the United Kingdom and the United

States. The Canada–US Free Trade Agreement (CUSTA), as articulated by Chiotti in Chapter 3, was promoted by the governments involved as enhancing market opportunities and bringing about prosperity. The result, however, has led to numerous disputes, and the potential downsizing of certain agri-food industries under the broader market pressures.

The diversification of the farm business

In the 1990s, the very nature and position of agricultural production in advanced economies, and the social relations upon which it is based, are subject to increased instability and debate. Munton *et al.*, in Chapter 4, argue that key aspects of family relations and market mechanisms are not necessarily structurally contradictory. It can also be pointed out that relative to subsumption, where fully commoditized production relations are apparent and the family farm has been completely absorbed, such cases are still in the minority.

World surpluses of traditional crops, changing patterns of consumption and the emergence of the entrepreneurial state have set the stage for farm diversification. Deslauriers *et al.*, in Chapter 5, substantiate how enterprise and marketing diversification is often associated with adaptive responses to the changing market for agricultural produce, particularly in the urban environment (see p. 7). This diversification cannot only contribute to a dynamic stability of agriculture in such areas, but also to the sustainability of the whole agricultural production system. Napton, in Chapter 6, summarizes how, in the United States, rural entrepreneurial diversifiers and industrial commodity diversifiers are approaching change in slightly different ways. Entrepreneurial diversifiers are basing change on experience and market analysis, whereas industrial diversifiers wait for the land grant university/agribusiness to prepare the market. Overall diversification is enhancing the environment and lowering risks and uncertainties associated with market fluctuations. In Chapter 7, Ilbery further suggests that two sets of factors will interact to affect the adoption of farm diversification: the characteristics of farm diversification itself and the characteristics of farms and farmers. As farm diversification becomes more widely appreciated, and knowledge of different schemes spread, resistance to adoption in contemporary systems will decrease. In addition, government programmes to encourage diversification can further enhance change.

Changing relationships between agriculture and environment

Despite a high level of interest in social and economic processes relative to changing rural systems, the role of environment is often ignored or believed to be irrelevant. Chapters 8, 9 and 10, by Roberts, Nellis and Hilts respectively, provide important insights into the re-emergence of the role of the environment in rural restructuring (see p. 4). As Roberts points out, the spatial pattern of changing rural systems is not only related to capital investment patterns but to natural patterns, because production technology combines nature and physical capital with the labour process. Further, simultaneous forces of differentiation and levelling generate a changing rural landscape reflective of natural processes that serves as a causal force in the development of politics of place.

The costs associated with environmental externalities relative to agriculture until recently were often ignored when assessing the rural system. Yet the transition of the rural system in the United States, for example, is strongly influenced by an accounting of these environmental costs. Nellis believes that the spectrum of environmental interest groups now involved in agriculture–environment policy issues, and emerging government programmes such as the 1990 Farm Bill, are creating a significant restructuring of the rural environment in the United States. This is further supported by the findings of Hilts in Canada. Despite the defensiveness of the agricultural industry and bureaucracy towards rural stewardship, there is increasing effort in Canada to ensure national and provincial policies that protect the environment.

The development of policy and programmes

A series of major policies and agreements have the potential significantly to modify the environment within which agriculture operates. Chapters 11 to 15 address a range of policy, trade agreement and planning issues that are creating a transition in the rural landscapes of Canada, the United Kingdom and the United States. Tarrant and Cobb, for example, in Chapter 11, demonstrate the convergence of agriculture and environmental policies in England. They argue that the objectives of agricultural and environmental policy came closer together with the introduction of the concept of extensification – that is, encouraging less intensive styles of farming. The initiative for extensification came from the European Community. This, along with other programmes (such as setaside) are only producing marginal success at best, due to confusion between EC policies and UK programmes.

The agriculture–environment issue (see p. 5) is further articulated in the fringe zone of urban areas. As Rickard points out in Chapter 12, however, the need for preservation of agricultural land on the urban fringe is still being debated. Questions of preservation, approaches at preservation, and viability of farming are still significantly variable among US state and local entities.

In contrast to the diversity of approaches on the urban fringe, in the United Kingdom there has been a convergence of policy and environmental protection relative to forest areas. Watkins, in Chapter 13, points out the growing recognition of the importance of ancient woodland and the increasing emphasis on lowland afforestation in the United Kingdom.

Rural lands utilized for research in industrialized countries also demonstrate the increased importance assigned to understanding sustainable rural land-use environmental systems. Harrington, in Chapter 14, summarizes the characteristics and issues associated with land areas controlled by public agencies and by non-governmental organizations in the United States devoted to research use. Agricultural/economic achievement, preservation of natural areas, military/energy technology development and extensive areas for global change studies are all purposes for which land is used in a research context.

As we look to the 1990s, rural systems will continue to change in response to further technical advances and reforms in agricultural policy in Canada, the United Kingdom and the United States. The resulting agricultural surpluses will lead in turn to surplus lands; Gilg examines this problem in Chapter 15 within the context of the policy environment and the recent planning response to land surpluses.

Sustainable agriculture as a policy option

The final part of Volume One considers the need to balance agricultural production against the damaging environmental consequences of modern farming methods. The concept of 'sustainable agriculture' is examined as a potential way of resolving the agriculture-environment dilemma. According to Pierce, in Chapter 16, most agricultural policy measures have pursued a small number of objectives to the detriment of more environmentally sound approaches. Pierce argues for policies that are more supportive of the environment yet promote sustainable agricultural systems. However, in Chapter 17, Bowler shows how 'sustainable' agriculture is interpreted as 'alternative agriculture' in the context of the European Community, with organic farming – the most sustainable of farming systems – given little policy support. Curry-Roper, in Chapter 18 is more optimistic of the outcome, at least for the USA, since a variety of popular movements have already emerged that support decentralization, an emphasis on community, and radical reconceptualizations of science.

Emerging Rural Systems: An Agricultural Environment for Change

Recent policy, programme and attitude changes in Canada, the United Kingdom and the United States, in response to a crisis in the rural system, have enhanced the potential for change in rural systems. Many of these changes reflect the need for longer-term sustainable agricultural systems so as to create more stable and diverse rural landscapes. Such systems need to offer lower costs of inputs, increased farm profits, maintenance of yields and reduced environmental impacts. The research reported in Volume One provides a basis for understanding the problems in creating such systems.

References

Bryant, C.R. (1989) Entrepreneurs in the rural environment. *Journal of Rural Studies* 5, 337–48.

Malassis, M. (1958) *Economie des exploitations agricoles: essai sur les structures et les resultats des exploitations agricoles de grande et de petite superficie*. Ecole Practique des Hautes Etudes, Paris.

Naisbitt, J. (1982) *Megatrends*. Warner Books, New York.

I

The Evolving Agri-food System

1

International Restructuring of the Agri-food Chain

Iain Wallace

It is easier to claim that the agri-food chain exists as a 'seed-to-consumer system' (Austin, 1974, p. 1), than to trace from start to finish the specific paths of causality and influence that give the concept coherence. Rather than providing a case study that could demonstrate a specific set of such linkages, this chapter identifies selective system-wide forces that are shaping the contemporary restructuring of the international agri-food industry. Although a growing proportion of the overall agri-food chain is only indirectly linked to agriculture, most developments in the markets for food and beverages ultimately have implications for farm-based production and the rural economy. Because large corporations play a dominant role in establishing the agricultural, manufacturing and distributive phases of functionally integrated food supply chains, they are the primary focus of research. It will become clear that such firms (some of which are farm businesses) conduct operations and transfer experience across national boundaries in ways that link processes of rural change in Europe and North America and even further afield.

The chapter has five sections. The first three involve a review of the restructuring forces in the international agri-food chain associated with: the acceleration of corporate concentration; the changing nature of consumer demand; and the moves on both sides of the North Atlantic to create more integrated continental markets. In the fourth section, some of the most immediately apparent rural and farm-level implications of this restructuring are identified, as well as the forms of response that it is eliciting. The final section provides a more detailed analysis of the impact of restructuring within Canada's agri-food sector.

Forces Driving International Restructuring

Corporate concentration

The 1980s, despite recession at the beginning of the decade and a stock market crash in 1987, witnessed not only an accelerated internationalization of capital but the increasing divorce of financial flows and gains from the 'real' economy of goods and services (Drucker, 1986). The frenzy of typically highly leveraged corporate takeovers and management buy-outs, which was one manifestation of this trend, included a disproportionate share of businesses in the agri-food sector. Table 1.1 lists some of the most prominent deals, the largest of which rank second only to mergers in the international oil industry in dollar value. This flurry of empire-building and corporate asset-shuffling can be explained within a number of theoretical frameworks. For instance, it clearly had (almost) everything to do with the realization of exchange value in the agri-food chain and very little to do with enhancing use values, in the sense of radically improving nutrition, food safety, security of supply, or even the range of consumer choice. Indeed, a common thread through the majority of transactions was the realization, or enhancement, of that 'fictitious' capital represented by the market value of a brand name. Cashing in on the icons of commodified consumption was especially evident in subsectors characterized by non-staple, 'life-style' products such as confectionery, alcoholic beverages and soft drinks, where the promotion of truly global brands is much more readily achieved than with foodstuffs more sensitive to national and regional variation in tastes; compare with Harvey's (1989, pp. 299–300) discussion of the culture of postmodern food systems.

Alternatively, branding can be viewed simply as a standard strategy for maintaining barriers to entry and engaging in non-price competition in 'mature', oligopolistic industries. Most subsectors of the food processing and retailing industry fit this description, although that is not to discount the significance of major technological developments such as the microwave oven or aseptic packaging leading to new rounds of product innovation. Seeing their prospects for further growth in established, especially domestic, markets increasingly constrained, processing firms have increasingly developed foreign markets, often by acquiring existing firms (as much for their brands and distribution systems as for their products and production facilities *per se*) with the intent of globalizing their strong brand names or adding complementary ones. Alternatively, as has been the case with the largest of all corporate reshuffles – the creation and then unbundling of RJR Nabisco – the value of the brand name has been cashed in by selling its various business elements to other firms active in particular subsectors in particular regional and continental markets, such as the sale of the European biscuit operations to BSN.

Table 1.1. Major changes of ownership in the international agri-food industry, late 1980s.

Target/purchaser	Sector	Date	Value US$ million[a]
Non-Canadian			
Rowntree(UK)/NESTLE(Switz)	Branded confectionery	88/07	4510
Kraft(USA)/PHILIP MORRIS(USA)	Branded foods	88/11	13123
Pillsbury(USA)/GRAND METROPOLITAN(UK)	Branded foods/restaurants	88/12	5774
RJR Nabisco(USA)/KKR(USA)	Tobacco/branded foods	89/02	24501
Holly Farms(USA)/TYSON FOODS(USA)	Branded poultry	89/06	1287
Nabisco(Europe)/BSN(France)	Branded biscuits/snacks	89/06	2500
Nabisco-snacks(UK)/PEPSI CO(USA)	Branded snacks	89/06	1350
Del Monte – processed foods (USA)/ MERRILL LYNCH & MANAGEMENT(USA)	Canned fruits and veg.	89/09	1475
Beatrice(USA)/CONAGRA(USA)	Branded packaged foods	90/08	1363
Jacobs Suchard(Switz)/PHILIP MORRIS(USA)	Coffee/confectionery	90/08	3800
Canadian			
Carling O'Keefe/ELDERS IXL(Austral.)	Beer	87/02	390
Maple Leaf Mills/HILLSDOWN(UK)	Milling and baking	87/08	360
Beatrice Canada/ONEX(Can)	Branded food/dairy/bisc.	87/11	316
Nabisco Canada – 20% not owned/ RJR NABISCO(USA)	Branded food/cereal/ canned fruit/veg.	88/03	180
Redpath – 50% not owned/TATE & LYLE(UK)	Sugar	89/05	286
Canada Packers – 56% control/ HILLSDOWN(UK)	Meat/branded meat prods/ veg. oils	90/04	556
Campbell Soup Canada – 29% not owned/ CAMPBELL SOUP(USA)	Canned soup/branded foods	91/07	154

[a] US$ values are calculated at prevailing exchange rates.
Source: Wood Gundy Inc., *Offer to Purchase* circular, Campbell Soup Company Ltd, May 1991: Toronto, Schedule 2; supplemented from author's press files.

This reaction to mature market conditions has not been confined to processors. It is also evident among supermarket chains, whose executives argue that internationalization is at least partly necessary to counter the power of the giant transnational food manufacturers (Wrigley, 1987). For instance, three major European retailers, Argyll (Safeway), Groupe Casino, and Koninklijke Ahold, operating respectively in the UK, France and the Netherlands, formed a new joint venture company, based in Luxembourg, by a mutual share exchange 'to exploit opportunities in marketing, distribution, purchasing, production and management information systems' (*Financial Times*, 1989d). A separate Swiss-based company, created to handle their joint buying, including that of fresh produce, also has retail chains from Denmark, Italy, Sweden and Switzerland as participants. In a different structure, common family control links the Tengelmann chains in Germany, North America (A&P), and the UK (minority holding in Gateway), and similarly the Weston chains in Canada, the USA, and Ireland. Branding among large retailers consists in developing a distinctive image and reputation, usually involving 'own label' goods, which compete with 'national' brands, as Loblaws has done in Canada with its 'President's Choice' and controversial 'Green' products (Brady, 1992), or, as in the case of Marks and Spencer, displace them entirely.

The upshot of this buying, selling and shuffling of corporate assets has been to intensify already high levels of corporate concentration in the off-farm components of the agri-food system. (The farm machinery subsector underwent its own rationalization in the early 1980s.) Not surprisingly, a fair number of mergers and proposed mergers, especially in basic commodities (for example, sugar in the UK, flour in Canada) and retailing (including the 'tied' outlets of British brewers) have been referred to national (or European Community) monopolies commissions, which have not left them unscathed. But despite their heavy reliance on debt instruments, the majority of these deals have proven to be financially successful. In identifying why, one can point to the strong and steady cash-flow that characterizes the food business; the potential for increasing added value if well-managed firms are prepared to keep abreast of changing consumer demands and to upgrade their products and service; and the benefits that many mergers have yielded in terms of reduced distribution and overhead costs. Rationalizations have also, of course, precipitated substantial redundancies and, in some US industries such as meat-packing and food retailing, notable reductions in unionization and wage levels. The implications of all these developments for the agricultural producer will be more readily appreciated if we first consider how they interact with consumer-driven changes to the agri-food system.

Changing consumer demand

As another characteristic of mature, consumer-oriented, oligopolistic industries is their heavy spending on advertising, it is appropriate to question the degree to which food customers' choices are really 'free', especially as ill-defined and often contested health claims have played an increasing role in product promotion. Nevertheless, there have been shifts in consumer purchasing that undoubtedly reflect wider social trends. The premium on convenience of food preparation, which stems from higher female labour-force participation rates, the growing proportion of meals taken outside the home, and demand for food at non-traditional times of day, among other factors, has supported rapid growth of the market for 'ready-to-serve' foods, particularly microwavable products. Heightened public concern over nutritional standards and food safety has been expressed in a number of trends, including the shift from red to white meat; the growing appeal of foods lower in fat, salt and calories and higher in fibre content; and a widespread willingness to pay more for fresh, 'natural', or organic produce than for processed foods or those associated with heavy chemical inputs. In particular, demand for canned foods has declined; it has been stable or increasing (slightly in North America, more rapidly in Europe) for frozen foods; but has expanded rapidly for fresh and refrigerated (chilled) items.

Successful retail chains have responded to and consolidated these trends, which have ramifications throughout the agri-food chain. Within the store, there has been a great increase in the space and prominence given to fresh fruits and vegetables, often at the expense (at least in North America) of that formerly given to meat. The fragmentation of product markets implied by the need to offer processed foods in their various 'low-calorie', 'low-sodium' formats, and the marketing of a wide variety of prepared meals or their non-generic constituents (for example, green beans in a range of formats) has greatly increased the pressure on shelf and freezer space. Together with the rising volume and scope of the stores' 'own brand' sales, this has reinforced the significance of brand leadership among food processors, for only strongly selling national brands are allotted, and can afford the shelving fees required to secure display space. With the exception of a few tropical products with vertically integrated supply chains, notably bananas and pineapples, branding has not become a dominant feature of fresh fruits and vegetables, nor are stocking rebates the norm; but growth in prepared fresh foods (for example, pre-assembled salads) provides an avenue for greater branding, whether by the retailer or a transnational firm, and the higher margins that go with it.

Upstream of the retailer, there have been major changes in the food distribution subsector. Heightened consumer demand for freshness and high quality means that the major retail chains, which in most cases are, or are affiliated to, their own wholesalers, have sought much closer control over product shipment. Compared with tinned, frozen, or even many fresh foods, refrigerated products have a short shelf-life; and the increased range of goods in that category (beyond the traditional dairy products) has added to the momentum for reconfigured distribution networks. Rather than having each supplier of everything from frozen shrimps to baked beans deliver to each store in its own vehicle, the retail chains have moved to large centralized warehouses, often run in the UK by specialist third-party firms, with a range of controlled-climate storage facilities, from which vehicles that are comparably equipped can make consolidated shipments to each store. In the process, elements of the 'just in time', inventory-eliminating, shipment practices widely adopted in the auto industry have been imposed on suppliers. Retail chains can gain significant savings in reduced handling, wastage and paperwork from such delivery networks, which in one respect displace part of the supplier's traditional distribution service. British firms with expertise in such logistics have already begun to expand into continental Europe, although the less concentrated structure of retailing there gives them less scope. Yet the branded-goods manufacturer and traditional wholesaler still need to maintain effective distribution channels to their non-chain customers or risk losing business volume.

The emergence of continental markets

A third driving force in the restructuring of the agri-food industry in Europe and North America has been the movement towards freer continent-wide trade. There are, of course, some important trans-Atlantic differences in the context in which this has been pursued. There is much greater similarity of consumer demand and channels for supplying it across the Canada/USA border than there is across the national boundaries of Europe. On the other hand, the restructuring of the food-processing industry in the European Community takes place against the background of the *Common Agricultural Policy*, whereas the 1988 Canada–United States Trade Agreement (CUSTA) glossed over the implications of Canada's regionally- and commodity-selective protectionism and its lack of congruence with the structures of agricultural support in the USA (see Chapters 2 and 3). On both continents, the removal of barriers to trade is leading to restructuring of the agri-food chain in ways that emphasize economies of scale in processing and the streamlining of distribution channels. Specialized plants serving extensive markets (such as the French frozen Mars Bar plant that supplies 16 European nations) are replacing multi-product plants whose shipments are more localized. The greater competitiveness that comes from loss of protected national markets is keenly felt, especially at plants whose output has been below capacity. Even consumer concern with freshness cannot always be guaranteed to work to the advantage of local suppliers, as transportation improvements and reduced border crossing delays assist long-distance shippers. Because of its corporate structure (high levels of foreign ownership) and more environmentally constrained agricultural supply conditions than its southern neighbour, Canada has been particularly exposed to the consequences of corporate rationalization in a freer trade context (see below).

Rural Consequences of Restructuring

The trends towards corporate concentration in both the food processing and the distribution and retailing subsectors mean that farmers face produce markets largely if not completely controlled by fewer and more powerful agents. This is not a new development in the industrialized nations, but it is an intensifying one. It has resulted in increasingly polarized farm sectors, wherein a small proportion of large and well-capitalized operations generate a dominant and growing share of marketed output (Gregor, 1982). The aggregate consequences of this process for rural areas have been at the root of many of the most challenging and contentious policy issues associated with the stalled Uruguay Round of GATT negotiations. But the forces of change are dynamic, and even if the general outline of trends in contemporary agricultural restructuring is clear, the mechanisms of agri-food chain integration are constantly undergoing modulation. Some of the more salient shifts are identified here.

With the streamlining of distribution infrastructures, often to the complete elimination of conventional wholesaling, retailers are forcing more managerial responsibility onto their suppliers. Given the growth in demand for fresh or chilled produce, this means increasingly that farmers as well as processing transnationals are required to 'deliver' – in more ways than one! A series of British case studies (*Financial Times*, 1989a,b,c) gives some insight into the demands and pressures imposed on produce suppliers and of the 'knowledge-intensity', not merely the capital intensity, that is required of them by leading-edge, 'interactive' buyers such as Marks and Spencer (M&S). For instance, Kent Salads (originating in two estate farms covering 800 ha) became a major supplier of prepacked salads following a fact-finding trip sponsored by M&S to California's iceberg lettuce industry. Importing the US field operations know-how to the UK did not provide satisfactory results until a lettuce variety more suited to Kentish growing conditions was identified. The transition from lettuce-grower to year-round value-added produce supplier involved Kent Salads in replicating its own stringent client relationship with M&S with grower-suppliers in France, Spain and Italy, to cover the British off-season and to source more exotic salad items (which also come from as far as Zambia).

In all of this, as in comparable producer–seller relations, the risk is borne almost entirely by Kent Salads. The level of management skill and willingness to live with decidedly asymmetrical power relations required to succeed in such relationships may not be novel in large parts of the United States, but it is now also becoming normative for commercial producers elsewhere.

At the other end of the scale, the consumer wariness of many aspects and products of conventional industrialized agriculture has been seen to provide new opportunities for small, 'alternative' producers, countering the trend to commercial marginalization of traditional farms. Undoubtedly, on both sides of the Atlantic, organic and low-input agriculture has assumed some importance, and its appeal may well be bolstered in the process of removing the price distortions embedded in current agricultural support policies (Clunies-Ross, 1990). It should be noted, however, that as the major retail chains respond to these expressions of consumer preference, they tend to shape a supply chain that has basically the same characteristics and demands on suppliers' managerial competence as their existing sourcing channels. It may be that regions with a supportive culture for alternative agriculture will nurture competitive commercial suppliers (for example, Organic Farm Foods, a Lampeter-based grower cooperative that also imports produce (*Financial Times*, 1990)), but the very different geographical scale of European and North American markets needs to be recognized. West Wales is much closer to southeast England than is central Saskatchewan to southern Ontario, and the logistics of supplying retail chains year-round are considerably easier. In this light, the emerging commercial zonation of the agri-food supply system suggested by Smith's (1984) 'vortex model' needs to be interpreted with sensitivity to the regional and continental context. Even assuming that demand for its output increases substantially, alternative agriculture is unlikely to make much difference to the rural prosperity of really peripheral regions, and its existing peri-urban practitioners, with their often volunteer-run marketing channels, are vulnerable to displacement by non-local suppliers delivering to the retail chains.

The tendency of the food-processing industry to concentrate an increasing proportion of its output in capital-intensive plants, most of them in metropolitan regions, will continue to reduce rural employment in this sector. That does not contradict the observation that food processing will remain more prominent in rural areas than manufacturing as a whole. Some market developments, such as the demand for prepared salads and other ready-to-eat foods for retail or catering trade outlets, actually involve increased labour-intensity. The rural consequences of continued corporate concentration in food retailing are subject to a number of other variables, including population growth/decline and levels of personal mobility. The continuing growth of out-of-town shopping centres, or stand-alone supermarket sites in Europe (and even still in North America), tends in aggregate to pull rural customers away from higher-priced local stores.

Examples of Restructuring in the Canadian Agri-food System

The international trends identified above have all been prominent in Canada in recent years. Different levels of exposure to 'world market' pressures between regions and agricultural subsectors, and a tendency for provincial governments to balkanize the national market in everything from eggs to beer have enhanced the difficulties and policy dilemmas associated with agri-food restructuring. As elsewhere, food processing is relatively more prominent in rural areas than manufacturing activity as a whole, so one question is the degree to which corporate restructuring is changing this spatial pattern. Because processors of fruits and vegetables have always been particularly raw-material oriented, changes in this sub-sector are bound to be felt predominantly in rural communities. Producers of more highly manufactured foodstuffs, or those dependent on imported commodities, have, conversely, been predominantly metropolitan. As this segment of the agri-food chain has grown disproportionately since 1945, even the accelerated pace of corporate restructuring may have few immediate rural consequences. Canadian analysts are particularly interested, however, in the influence of CUSTA on corporate rationalization in an increasingly integrated continental market.

The Campbell Soup Company

The experience of the Canadian branch of the Campbell Soup Company illustrates many of the pressures on foreign-owned processors and their suppliers. From 1931, when it opened a plant in Toronto, until 1983, Campbell was a wholly owned subsidiary of its US parent. Then 30% of the equity in the Canadian unit was sold to Canadian investors, with the rationale that a local management would be more responsive to the requirements and opportunities of the Canadian market. By that time the firm operated 10 manufacturing plants in four regions of Canada and owned three farms (Table 1.2). Under an energetic and 'visionary' Canadian president (Trueman, 1991), the Canadian branch subsequently adjusted, comprehensively and profitably, to changing consumer demands. This involved closure or sale of peripheral, low value-added businesses, such as vinegar and apple juice, but capitalization on expertise in handling fresh and refrigerated foods (initially developed in the corporate mushroom farms that feed into soup production). Purchase of a Toronto-area supplier of delicatessen food in 1989 was seen as providing 'a laboratory' for product development in that growth sector. The core soup business (nearly half the Canadian market in the mid-1980s) was consolidated the same year by the purchase of the leading Quebec brand (Habitant) from Catelli (which was being sold off by the large beer and dairy firm, Labatt, as a non-core subsidiary).

Table 1.2. Campbell Soup Company, Canadian operations 1984 and subsequently.

Plant location	Products	History
Toronto	Canned foods	Expansion in 1990s
Toronto	Juices/condiments	Closed 1985
Chatham, Ont.	Juices/veg. processing	Production consolidated
Listowel, Ont.	Frozen foods	Changed labour process
Thornbury, Ont.	Juices/condiments	Closed 1985
St Mary's, Ont.	Poultry processing	
Wolfeville, NS	Juices/condiments	Sold 1988
Saskatoon	Condiments	Sold 1987
Portage la Prairie, Man.	Canned foods	Closed 1991
Montreal	Italian specialities	
Farm location		
Listowel, Ont.	Poultry	
Wellington, Ont.	Mushrooms	Sold 1990
Portage La Prairie, Man.	Mushrooms	Sold 1990

Sources: Campbell Soup Company Ltd., Toronto, *Annual Report 1984*; author's press files.

However, within a few months of the Habitant purchase, Campbell announced that its 30-year-old plant in Portage la Prairie, Manitoba, would close in January 1991, with a loss of 168 jobs. (The same day, the US parent announced four plant closures, involving 2800 jobs.) The company argued that it had too much Canadian production capacity in what was a mature product market and, in the context of continental free trade, pointed to the unit cost differential between its 2.5 million cases/year in the Portage plant and its new 50 million cases/year plant in Napoleon, Ohio. The future of the firm's Toronto factory, which in 1989, in classic branch-plant style, turned out 480 different products (including soups), was to be secured by its specialization in 'low-volume' lines for the continental (especially northeastern US) market, but in quantities representing a threefold increase in total output. At the firm's frozen foods plant in Listowel, Ontario, recognition that survival depended on cutting the 25–40% production cost premium over equivalent US plants prompted a revolution in labour organization, towards worker self-management teams, as well as more active and stringent involvement with input suppliers. Nevertheless, the operational integration of Campbell

Soup on a North American basis soon led to the parent's decision to buy out the Canadian shareholders, to achieve unified corporate control, leaving the Canadian plants and their suppliers more remote from the firm's decision-making processes.

Campbell's restructuring involved geographical consolidation at two scales: pulling back from locations in the national periphery (Manitoba and Nova Scotia) to concentrate operations in Ontario, and within Ontario to consolidate on plants in larger centres. This pattern should not be over-generalized, as it partly reflects the firm's specific products and past investment decisions. However, it is the inverse of the spatial restructuring of the Canadian meat-packing industry, which is moving from a metropolitan focus in central Canada to a small town focus in western Canada, very much along the lines of the restructured US industry (see Chapter 3 and Broadway and Ward, 1990). But it illustrates wider trends that are leading to the closure of Canadian branch plants, primarily because of their uneconomic scale compared with modern factories south of the border, and also, despite their small size, because of their grossly underutilized capacity. Some of this inefficiency undoubtedly reflects structures of foreign ownership that have prevented Canadian plants from producing for export (especially US) markets; yet rationalization has not been confined to foreign-owned firms, and one must conclude that there have simply been too many individual food-processing plants, many of them seasonal operations only, for viability in the market conditions of the 1990s. Just as the distribution of Prairie country elevators has been steadily thinning out, despite legislation designed to retard the process, it should be no surprise that a similar response to improved local transportation is evident in, for instance, fruit and vegetable growing areas such as southwestern Ontario. The diminished popularity of canned products has aggravated the excess capacity, and it is noticeable that some of the canneries recently closed are 60 years old and over.

Overarching all these factors are the problems, which very many Canadian food processors face, of matching not just the scale but the input costs of US plants. Supply management in areas such as dairy products and poultry raises Canadian raw product prices above prevailing US prices (see Chapter 3), and for other farm products, where marketing structures have less influence on pricing, the length of Canada's supply season compared with that of more southerly growers often involves cost penalties for processors. Although the immediate post-CUSTA response of US firms with a major presence in Canada, such as Campbell and Heinz, has been to seek viability through cost-cutting, the threat to move processing machinery, and with it the raw-product supply chain, south of the border has been openly stated. Even McCain Foods, a major Canadian-based processor that opposed the CUSTA, anticipating just such results, has protected its interests by buying US food processors and thereby access to US input supply chains.

McCain Foods Limited

McCain provides a rare example of an agri-food transnational corporation based in a rural periphery: indeed, its global headquarters are located in Florenceville, New Brunswick, a settlement of 700 people. It has retained its structure as a private, family-run firm (more akin to large food firms in continental Europe than Britain or North America), despite having grown in 35 years to attain global sales in excess of $2 billion from over 40 plants worldwide. In the early 1980s, McCain began to purchase shares in Canada Packers, then family-controlled, but these were subsequently sold when its buy-out attempt was strongly resisted by the dominant shareholders (who held on until the Hillsdown purchase in 1990, see Chapter 3). Whether the firm's dynamic internal growth and continued expansion into new markets, such as Eastern Europe, will be sufficient to maintain its independence, or simply increase its appeal as a takeover target by firms in the Nestle or Unilever league, remains to be seen (*The Financial Post*, 1990).

The growth of McCain Foods has been fuelled by a combination of attention to changing consumer demand and a strategic diversification in the use of assets. Its transformation, from a shipper of fresh potatoes to a processor of frozen French fries, caught and helped to accelerate the tide of convenience food innovation in the late 1950s. The firm's remoteness encouraged the internalization of functions through vertical integration, leading to the creation of subsidiaries involved in activities ranging from land ownership and the production of potato-harvesting machinery to trucking and produce storage. The infrastructure of frozen food storage and distribution has provided the basis for diversification into frozen vegetables, fruit juices, pizzas (and so into cheese) and other prepared meals. Geographical expansion has given the firm market leadership in frozen French fries in Canada, the UK and western Europe, with additional substantial sales in the USA and Australia.

In its home region of the St John Valley, New Brunswick, McCain is essentially a monopsonist of the potato crop, and its relations with contract growers have inevitably been open to accusations of exploitation (Murphy, 1987; Glover and Kusterer, 1990). The capital requirements, harvesting technologies and preferred varieties for processing have effectively forced farmers to choose between a McCain contract or growing for the fresh potato market, so reducing their options. Those farmers best able to satisfy the company's requirements have certainly prospered most, but Glover and Kusterer (1990, p. 93) conclude that McCain's net effect in the region has been positive for all growers: 'without a processing firm, potato farming would likely not be viable in New Brunswick today' and the related agri-food employment would be much reduced.

McCain's product mix and international operations have prompted corporate responses to the Canada-United States Trade Agreement that may appear schizophrenic compared with the stance of other firms in the Canadian agri-food sector, but which are internally coherent. Having opposed the CUSTA, unlike transnational food firms less-rooted in Canada, McCain has not hesitated to act rationally in the wake of its implementation. Because frozen pizza production costs are 18% higher in eastern Canada than in the northeastern United States, mainly because of the higher-priced cheese from Canada's supply-managed dairy sector, yet US pizzas will be able to enter Canada duty free after 1993, McCain has bought a New Jersey pizza factory, which could supply its Canadian market if necessary. And its prior opposition to the trade agreement has not prevented McCain from hiring Canada's former chief negotiator to spearhead its fight against proposed expansion by a rival potato processor, Cavendish Farms (part of the K.C. Irving group of companies). Cavendish's proposals, first to build a new frozen French fry plant in Prince Edward Island, and then one across the road from McCain in Grand Falls, New Brunswick, both relying on substantial government subsidies, have been heavily attacked by McCain for threatening all Canadian frozen French fry exports to the United States with countervailing duties. Cavendish cancelled both proposed plants in late 1991, citing deteriorating market conditions, but not before McCain had carried out its tactic of facing-off Cavendish by building an unsubsidized plant in Prince Edward Island (PEI) at a time when its existing plants in New Brunswick were operating at only two-thirds capacity (*The Globe and Mail*, 1991).

Conclusion

This chapter has outlined, and in a few instances detailed, chains of causation that are translating international corporate restructuring in the agri-food sector into changes taking place in rural regions, where most foodstuffs originate. These forces, essentially imposed from 'above' and 'outside', still allow some freedom of response by those rural actors able to adjust to a more demanding and turbulent business environment.

References

Austin, J. E. (1974) *Agribusiness in Latin America*. Praeger, New York.
Brady, D. (1992) The names game. *Macleans* 3 February, 64–6.

Broadway, M.J. and Ward, T. (1990) Recent changes in the structure and location of the US meatpacking industry. *Geography* 75, 76–9.

Clunies-Ross, T. (1990) Organic food: swimming against the tide? In: Marsden, T. and Little, J. (eds), *Political, Social and Economic Perspectives on the International Food System*. Avebury, Aldershot, pp. 200–14.

Drucker, P.F. (1986) The changed world economy. *Foreign Affairs* 20, 768–91.

The Financial Post (1990) McCain . . . chip wagon to the world. 15 January.

Financial Times (1989a) Profits in store from contented porkers. 25 May.

Financial Times (1989b) Ready-made profits at the salad counter. 22 June.

Financial Times (1989c) Harnessing farming profits to high street power. 11 August.

Financial Times (1989d) Three European grocery chains to swap shares in joint venture. 5 September.

Financial Times (1990) A race to keep up with demand. 17 September.

The Globe and Mail (1991) Spud spat sputters. 23 November.

Glover, D. and Kusterer, K. (1990) McCain Foods, Canada: the political economy of monopoly. In: Glover, D. and Kusterer, K. (eds), *Small Farmers, Big Business: Contract Farming and Rural Development*. Macmillan, Basingstoke, pp. 73–93.

Gregor, H.F. (1982) Large-scale farming as a cultural dilemma in US rural development – the role of capital. *Geoforum* 13, 1–10.

Harvey, D. (1989) *The Condition of Postmodernity*. Basil Blackwell, Oxford.

Murphy, T. (1987) Potato capitalism: McCain and industrial farming in New Brunswick. In: Burrill, G. and McKay, I. (eds), *People, Resources, and Power: Critical Perspectives on Underdevelopment and Primary Industries in the Atlantic Region*. Acadiensis Press, Fredericton, pp. 19–29.

Smith, W. (1984) The 'vortex model' and the changing agricultural landscape of Quebec. *Canadian Geographer* 28, 358–72.

Trueman, W. (1991) Alternate visions: resurrecting Campbell Canada. *Canadian Business* March, 29–33.

Wrigley, N. (1987) The concentration of capital in UK grocery retailing. *Environment and Planning A* 19, 1283–8.

2

THE RESTRUCTURING OF AGRICULTURE: THE CANADIAN EXAMPLE

Michael Troughton

This chapter attempts to identify and describe critical changes affecting contemporary Canadian agriculture. Although emphasis is on change at the farm level, it addresses the 'total (Canadian) agricultural system' first outlined by the 1969 Federal Task Force, which places farming within the context of agribusiness, the activities of governments and the preferences of consumers (Federal Task Force, 1969, p.5). The underlying premise is that, whereas for 200 years Canadian agriculture has been a dynamic system expanding continuously until the 1950s, yet evidencing continuous adjustment including contraction, the most recent ongoing changes, although part of a long sequence, represent fundamental system restructuring. Whereas previous adjustments and rationalization at the farm level were occasioned by combinations of economic, technological and institutional circumstances, national and international, nevertheless agriculture remained a distinct production system and part of a wider rural system. Restructuring is destroying both the distinct structure of production and agriculture's rural function. Agricultural production is now concentrated on a minority of farms, increasingly integrated within an industrialized agriculture and food (agri-food) system, in which control has passed to agribusiness and government. The remaining majority of farms are economically marginalized within an agrarian system in which decoupling of agriculture from rural society and environment is well-advanced.

The chapter is divided into three parts. First, the restructuring process is described. This process, the momentum of which grew after 1950 through increased dependency on secondary inputs, links to the processing sector and increased government involvement. These forces promoted modernization and selective 'industrialization' of the system. Application of an industrial model has been the major factor creating the minority of industrialized production units and furthering the decoupling processes. Government policies, while professing to serve farmer interests, have generally aided the process.

The results of restructuring have been drastic, and operational and structural changes at the farm level are detailed in Part Two. Of critical importance are distinctions between the minority of 'industrial' and the majority of more traditional farm units. The former comprise less than 40% of total farms but produce over 85% of gross agricultural output and revenue. This situation is marked by significant differences of farm tenure and organization, in the scale and intensity of application of capital and in levels of specialization. In addition, important linkages have been established between key production and processing sectors with governments acting generally to facilitate this situation. Finally, in Part Three, attention is given to significant impacts of restructuring on the rural system, namely its physical environment and broader socioeconomic structure. Environmental impacts are serious enough to call into question the ecological sustainability of the system, whereas socioeconomic impacts include the creation of widespread marginality in both farm and rural service sectors. In conclusion, the vulnerability of a restructured Canadian agriculture is suggested within the wider context of international trade negotiations.

The Restructuring Process: Mechanization to Industrialization

In a modern economy and society, such as Canada by 1940, agriculture is both in relative decline and subject to a range of influences from within and outside the farm and broader agricultural system. Post-war agricultural adjustment or initial restructuring took place in response to two primary situations: the demand for and attraction to labour from non-agricultural sectors, and the internal shift that centred on rapid mechanization as a means to both substitute for labour and to increase agricultural productivity. In many parts of eastern Canada, large numbers of farmers were unable to modernize because they occupied land that could not be effectively mechanized or fertilized, while many others saw greater opportunities in off-farm employment. The result was a massive migration from farm to urban areas, resulting in widespread farm abandonment in marginal areas across eastern Canada. Even core areas of southern Quebec and Ontario experienced major population losses, but in these areas this hastened increased mechanization and modest increases in farm size. Overwhelming losses in peripheral regions began largely unnoticed, but were well-documented by the 1961 Census as being associated with rural farm and non-farm poverty. This prompted major government intervention through policies and programmes for agricultural rehabilitation and regional economic development (Buckley and Tihanyi, 1967). These measures, although largely ineffectual, established government involvement (federal and provincial) in the restructuring process.

The mechanization of agriculture

Mechanization, while proceeding apace in eastern Canada, especially in the emerging cash crop areas of southwestern Ontario, had by far its greatest impact in western Canada. Horse power on Canadian farms peaked in the 1920s, as a result of the huge numbers required to plough and harvest Prairie grain. By 1950 scarcely any horses remained, replaced by tractors, swathers and combines of increasing capacity which contributed to a very rapid increase in average size of Prairie farms (164 ha in 1941 to 250 ha in 1961). This shift, which involved major increases in improved and cropland areas, was on a steady farmland base and therefore meant a major reduction in farm numbers (−30%, 1941–61). Some of this reflected units already abandoned in the Depression, but there were also major out-migrations to both Prairie cities and further west. Because of mechanization, the farm labour force experienced massive decline; for Canada as a whole, between 1939 and 1961, farm labour (including operators) declined by almost 50%, while farm numbers fell by 35%.

Although this adjustment facilitated the modernization of farming, including increases in productivity and total output, it severely disrupted previous balances within the agricultural system. The sectors that benefitted most were agribusiness input supply and output processing and distribution. The former not only manufactured and supplied the increasing amounts of machinery, but was also involved in the rapid development of the fertilizer and agricultural-chemical industries. Farming rapidly became more dependent upon these secondary inputs, the control of which became established in a small number of major manufacturing companies. Increased input investment also involved government (Farm Credit Corporation) and private financial institutions. Agricultural processing continued its transition from small local, to large centralized facilities, including sectors such as dairying, and food distribution saw the emergence of supermarket chains (see Chapter 1). The combined power of agribusiness contributed to the widely observed 'cost-price squeeze' that effected even efficient farm operators (Mitchell, 1975).

The effects of the 'cost-price squeeze' and pressures on farm income resulted in further restructuring among the remaining farms, particularly in eastern Canada. Whereas the majority attempted to make the necessary investments and sought success through increased agricultural output and farm receipts, a sizeable minority were either unwilling or unable to invest totally in farming and adopted a part-time mode. For some, the combination of farm and off-farm employment and income was a stepping-stone out of agriculture, for others it was a persistent condition (Fuller and Mage, 1976). In the Prairies, there was continual pressure from both railways and elevator companies to rationalize the system, but farmers generally retained government support against closures that threatened the fabric of rural communities.

In retrospect, the 1960s were a critical decade. Despite recognition of continuing structural problems, it seemed that adjustments at the farm level had achieved the shift to an efficient production system, and drastic farm reductions of two post-war decades had produced a more streamlined but stable farming system, supported but not dominated by agribusiness, with governments playing a major but generally supportive role. In fact, a new set of relationships was emerging and events were in train leading to more fundamental restructuring in a wider systems context. A new paradigm was emerging, namely, the shift from agriculture combining a 'way of getting a living' with a 'way of life', to that of 'agriculture as a business enterprise'. This latter had long been the credo of agribusiness, but now dominated other influential elements – researchers, economists, and government agricultural departments – and was promoted as the basis of successful farming. The business enterprise paradigm was expressed through application of a new model of agricultural activity, namely, agricultural industrialization (Troughton, 1986).

Agricultural industrialization

The industrial model is driven by forces beyond the farm gate. Agriculture functions as an 'input-output' or 'assembly line' sequence, in which the actual production (farming) utilizes and is dependent upon inputs from agribusiness (machinery, fertilizers, capital, etc.) and furnishes output to the specifications of agricultural processers and distributors, who control the market for agricultural products. Decisions are made primarily by agribusiness, supported by government policy. The latter, although often presented as benefiting farmers, is increasingly conditioned by the tenets of industrialization, namely, that agriculture should achieve economic efficiency expressed through higher levels of land and labour productivity and lower unit costs of production. The primary route is through application of increased inputs of agricultural science and technology, including, where appropriate, 'factory farm' and/or specialized assembly line approaches, linked to specified market demands. Industrial model system goals combine agribusiness interest in profit with government, and consumer interests in cheap, quality food supplies. To farmers it offers the attractions of an efficient farm business operation with commensurate financial rewards. In practice, it has to contend with variations inherent in a complex economic sector – fluctuation in demand and supply, problems of national and international policy. In the case of agriculture other major problems stem from limitations in the model. First, industrialization applies to only some farms and takes little account of those operating in a more traditional context; second, it tends to ignore the inherent relationships between farming and environment; based on short-term financial goals, it employs an economic versus an ecological synthesis (Schnaiberg, 1975); and third, by excluding the social 'way of life' value in favour of business and institutional arrangements, it results in decoupling of agriculture from its rural socioeconomic environment.

Industrialized agriculture began to emerge comprehensively in Canada in the 1960s. As in the past, change was strongly influenced by US innovations, particularly new inputs, which began to change the nature of livestock production. New varieties of hybrid corn (maize) were introduced into southern Ontario (and later Quebec), which, combined with inputs of fertilizers and chemicals plus investment in tile drainage, vastly increased both unit area and per farm output of feed. Combined with soyabeans, corn provided the basis for intensive livestock systems based on housed flocks and herds. The shift began in the 1950s with poultry, and quickly resulted in battery production of eggs and 'factory farming' of broiler chickens and turkeys. This type of production met all the criteria for industrial agriculture, using new inputs to vastly increase the scale and lower the unit cost of ouput. The result was a rapidly reduced number of poultry farms, with those that remained being large-scale, highly specialized, capital-intensive units tied to agribusiness suppliers or processors. Poultry remains the most concentrated livestock enterprise, but similar increases in scale and intensity of production developed in hog (pig) and cattle production. Industrialization saw mixed livestock give way to specialized pig, beef and dairy farming. Despite the reduction in farm numbers, intensification quickly led to overproduction and depressed product prices, which, in turn, brought calls for an increasing role for governments.

Agricultural policy in Canada has largely pursued an overall goal of stabilization (Fulton, 1987), with policy measures aimed at maintaining farm incomes, including some measure of control over output and prices. Despite the general orientation, however, policy tends to be reactive to particular circumstances, including sectoral overproduction and declining farm income. In the early 1960s, emphasis was on attempts to stabilize agricultural land use. It rapidly became apparent that farm incomes were the problem, but factors affecting income reflected the emerging structure of specialized production, restructuring of agribusiness and changing domestic and international demand for Canadian agricultural items. Responses in the 1960s reflected this complexity.

In Ontario, the Special Committee on Farm Income (1969) concluded that only 40% of farms remaining in 1966 were needed to meet domestic demand, but that income viability could only be achieved by large specialized farms and that only 10% of enterprises met viability criteria. Another enquiry examined problems of production, organization and income in Ontario's dairy industry (Hennessey, 1965). It recommended drastic reorganization, including production to fluid standards and to meet dairy industry demand, and with a quota-based, supply management system to maintain farm income. Both provincial and federal governments acted on this blueprint, establishing the Ontario Milk Marketing Board (OMMB) and the Canadian Dairy Commission (CDC) respectively. The OMMB applied the model directly to Ontario, whereas the CDC implemented a national supply management system based on provincial quotas for milk production. The result has been a massive reduction of farms with a dairy component (150 000 in 1966 to 34 000 in 1986).

The Prairies, too, witnessed several initiatives, including partially successful attempts by railway companies to reduce their systems. Wheat farmers benefited from sales by the Canadian Wheat Board (CWB) to China and the USSR, but by the late 1960s demand was down, prompting a scenario for downsizing the system, including 30% fewer farms (Craddock, 1970), and an actual reduction in the area planted to wheat in 1970 (LIFT programme). By 1971–72, global demand led to a complete reversal of that policy but market problems encouraged a minority of large growers (Palliser Group) in efforts to weaken the cooperative Pools, and end the CWB monopoly; they established links with large private competitors, including US-based multinational suppliers and grain companies (Mitchell, 1975).

The Federal Task Force (1969) summarized the evolving situation and presented the integrated 'industrial' model as its blueprint for the future, including an expanded role for governments. It presented a series of economic goals leading to a streamlined, efficient system based on a series of commodity sectors. The latter emphasized the increased specialization of both production and processing. Most significantly, the Report outlined an institutional structure that explicitly linked government, agribusiness and farmer organizations, recommending a more integrated system, specifically identifying marketing boards as a mechanism for stabilization of production and farm income, and promoting the dairy supply management model. The Report also stressed the need to streamline production and transportation in the western grain system. Finally, while recognizing the low-income sector, it rejected any social maintenance policy.

While specific conditions changed quite radically, the Report formed a basis for federal policy in the 1970s, resulting in the extension of supply management to various sectors of the poultry industry and measures to rationalize Prairie rail and elevator systems and western grain freight rates (Hall, 1977; Gilson, 1982). However, while the federal government sought to implement farm stabilization policies, the measures tended to favour agribusiness, consolidating its entrenched oligopolistic position by guaranteeing markets for inputs and/or product supplies (Warnock, 1978;

Kneen, 1989; Troughton, 1989). Farmers continued to feel the 'cost-price squeeze'. The paradox was revealed in events over the 1975–85 period. Initially, increasing global demand led to higher commodity prices; this, coupled with inflation, led to large increases in gross farm receipts and the values of capital investment, especially land. Many farmers were persuaded to expand (i.e. industrialize), despite rising input costs and very high interest rates. In the ensuing recession, as commodity prices and land values fell much faster than expenses or interest rates, many farmers went bankrupt and many of the 'most efficient' producers were in financial trouble. Although agribusiness inputs were affected, most major suppliers diversified, and processors benefited from cheap prices. Meanwhile, governments had to provide massive subsidies to maintain farming activity. As global overproduction (relative to effective demand) and price wars have maintained low prices, the current situation is of subsidies comprising the majority of net farm income, particularly of the minority of 'industrialized' farmers producing the majority of output. Without subsidies, real farm incomes in the Prairies would be lower than during the 1930s Depression (Fulton *et al.*, 1989).

In summary, the last 30 years have seen a radical shift in conditions surrounding agricultural production activity. Whereas an earlier phase of 'mechanization' was dominated by responses to changes taking place within the farm sector, the later phase of 'industrialization' finds the farm sector impacted by and attempting to react to system changes dominated by agribusiness objectives and government policies. This is the joint product of a system becoming more industrialized and integrated and in which larger structural elements assume a natural dominance, but this is exacerbated because farming has become increasingly fragmented into specialized production sectors and its productive and residual parts.

Restructuring of Canadian Farming During the Last 30 Years

During its period of agricultural industrialization Canadian agriculture has experienced tremendous changes in structure and performance. The former includes continuing decline in numbers of farms, increases in average farm size and significant changes in patterns of tenure and organization. However, all this is set within the context whereby application of the industrial model to farming has involved increases in scale of operation, capital intensity and farm-level specialization on industrialized farms, with a widening gap between them and the remainder in terms of input and output measures.

Polarization in agriculture

Between 1961 and 1986 farm numbers in Canada declined by 40% from 480 000 to 293 000. While the annual rate of decline dropped from 2.3% in the 1961–71 decade to 1.3% in 1971–81, it rose again in the 1981–86 period to 1.6%. The largest declines were in eastern Canada, especially in areas where marginal physical and economic situations were most marked (Nova Scotia 66%, New Brunswick 70%, Quebec 57%). Although regional decline in the Prairies was only 29%, the absolute loss of 62 000 farms represented over one-third of the national decline. Farm sizes continued to increase in both major regions from averages of 60–80 ha to 77–115 ha in eastern provinces, and from 250 ha to 355 ha in the Prairies (420 ha in Saskatchewan), where there was a marked increase in farms in the over 650 ha census category, and evidence of polarization between larger groups (over 260 ha) and the 'smaller' farms. In eastern Canada, increase in size was linked more to particular enterprises (cash cropping, beef and dairying) with polarization related to an increase in farms below 25 ha, many in the hobby farm and non-commercial census categories.

While the overwhelming majority of Canadian farms continue as family operations, with less than 1% non-family corporations, significant tenurial changes have occurred. One response to uncertainty and a business orientation has been an increase in partnerships and family incorporations. Together, these have risen from 10% in 1971 to 17% in 1986. These 'non-traditional' tenure forms are much more common among highly capitalized and high gross income farms. Of farms with a capital value of over C$500 000 the level is 40% and rises to 64% of farms valued at over C$1.5 million. Similarly for farms with over C$100 000 gross farm receipts, 45% are in the 'non-traditional' category, rising to 76% of farms that earn over C$500 000. A more widespread adaptation, particularly apparent in connection with increasing area size, has been to increase rental of part of the farm base. While the proportion of full tenant farms has hardly risen, 'part-owner, part-tenant' increased from 21% in 1961 to 34% of all farms in 1986, with farm numbers remaining at approximately 100 000. These farms increased their share of farmland from 50% in 1971 to 60% in 1986. Their concentration is greatest in the Prairies where farmland rental reaches nearly 60% in some areas, but a similar trend is discernible in the industrialized cash crop regions of eastern Canada, including areas around the major metropolitan centres.

In terms of financial measures of both input and output, the pattern is of increasing polarization. By 1986, farms grossing over C$50 000 represented 38.6% of all farms but accounted for 86.4% of total output receipts; those grossing over C$100 000 (19.9%) represented 67.6% output, whereas the 45% of farms grossing under C$25 000 accounted for only 5.4% output. A recent study based on constant 1975 dollars indicates that, although the number of farms in the C$50 000 plus class (C$70 300 + in 1986 C$) has risen from 9.2% in 1971 to 29% in 1986, contributing 44.5% to gross receipts in 1971 and 78.4% in 1986, the number of remaining farms, although falling from 90 to 70%, has seen a decline of receipts from 55.5 to 21.6% (Brinkman, 1988).

Closely related to receipts are levels of capitalization; both fixed and operating amounts have risen rapidly, but polarization is again marked. Average farm capital values in 1986 ranged from C$154 000 per farm for those under C$25 000 gross sales, to C$2.1 m. for farms with sales of over C$500 000. Farms with capital values of over C$500 000 represented 21% of all farms, but accounted for 55.6% of fixed capital value, 56.4% of operating expenses and 57.4% of gross receipts. In contrast, the 41.6% of farms with capital values under C$200 000 accounted for only 12.8% of capital value, 11.4% of expenses, and 10.8% of receipts.

Other characteristics that emphasize polarization include measures of farm and off-farm labour, and concentration within specific enterprise types; the 21% highest capital value farms and the 19.9% with highest receipts, employ 57.8% and 66.8% of the farm labour, respectively; whereas the 41–46% farms in the lowest categories employ only 9–10% of the labour. In contrast, low gross income farm operators account for over 80% of off-farm (part-time) employment. Farms' enterprise categories reveal that from 1961 to 1986 the proportion of farms in 'mixed' categories has decreased from 11.8% to 5.6%. The largest declines by specialized enterprise types were in areas of supply management, notably dairying (−57%), and poultry (−53%), whereas less- or non-regulated enterprises types (small grains excluding wheat, beef, and miscellaneous specialities) actually increased. On the other hand, dairy and poultry have by far the highest proportions of farms in the highest capital-value and income-per-farm classes, with the opposite for beef cattle and, recently, wheat, reflecting severely depressed commodity prices. Despite the range of production there are signs of increasing concentration by a small minority in each enterprise category. This is most apparent for poultry, where the top 5% of farms by commodity sales account for 75% of livestock numbers and gross receipts. Other areas of high concentration include vegetables (40.5%), hogs (38.9%) and fruit (38.5%). Once again, however, dairy has the largest concentration in the top 25% in terms of gross farm sales (51%), followed by hogs and poultry (Brinkman, 1988). The linkage with supply management and contracts with domestic processors seem to be underlying factors.

Changes in farm incomes

Maintenance of price and access to markets are more important in terms of net farm income. Low commodity prices, especially export grains and oilseeds (wheat, barley, canola) domestic feedgrains (corn), and red meat (especially beef) mean that even industrialized producers have experienced declining net farm incomes and asset values. Although the aggregate net farm income of farm operators increased from C$1.4 to C$5.5 billion from 1971 to 1986, recalculation in 1987 dollars indicates that whereas aggregate income rose from C$4.7 billion in 1971 to C$9.5 billion in 1975, it fell to C$3.1 billion in 1983 and only recovered to C$5.4 billion by 1987. Furthermore, net income since 1979 has been markedly affected by government payments which rose from C$800 million to C$3.4 billion (1979–87); this means that actual or net 'market' income declined (in 1987 constant C$) from C$8.6 billion in 1973 to C$2.0 billion in 1987 (Brinkman, 1988). Although most subsidy payments go to major producers, support levels have scarcely covered costs of production, especially among grain and beef producers. The decline in net income was paralleled by the value of fixed assets. In actual value, farm capital declined by C$29 billion between 1982 and 1987, whereas debt increased by C$18 billion (to C$22.9), causing equity levels to fall from C$112 to C$82 billion (Brinkman, 1988). Farmland and building values (in constant 1987 C$) actually declined by C$68.5 billion between 1982 and 1987 (C$11 billion per year). During this severe financial downturn the only enterprises cushioned were in supply managed sectors where prices are indexed to costs of production.

A review of organizational and economic restructuring highlights the major shift in concentration of meaningful economic activity within the Canadian farm sector to a minority of industrialized farm operations. However, the use of technology and economically efficient production provides no guarantee of financial security; this is more a function of institutional arrangements that protect individual producers from a highly competitive marketplace, inelasticity of demand and oligopolistic agribusiness.

Restructuring as Decoupling

Agricultural restructuring centres on the changing characteristics of farms, but is taking place within the wider contexts of agrarian and rural systems. Restructuring, particularly the application of the industrialization model, has largely negative impacts on rural systems. The impacts fall into two broad classes: effects on the physical environment and effects on rural socioeconomic systems. Despite a number of distinct problems, the most important is separation or decoupling of industrialized farms from their overall setting, which threatens the viability of the total agricultural system.

Industrial agriculture involves an increased dependence on non-renewable resource inputs, and uniformity of crop and livestock components, each of which impact on the physical base and on which agricultural sustainability ultimately depends. Increased mechanization, large-scale monoculture, 'improvements' in drainage, amongst other factors, increase levels of soil erosion and loss of nutrients. Attempts to counter losses by fertilization and chemical application have led to increased run-off and water pollution. Another source of pollution is the huge concentrations of livestock. Landscape amenity is destroyed by mechanical and chemically maintained 'prairies'. Some authorities claim losses in quality of output (Hill, 1985), and there is an emerging consensus on the need for less intensive more conservationist approaches. However, a return to 'stewardship' is most difficult for those on the industrial 'treadmill', for whom de-intensification is operationally at odds with the technological and economic elements of the industrial model. Consequently, although there is interest in alternative, even organic production, at present individual farmers find it difficult to adopt, and governments in Canada have not yet developed policies or incentives. Agribusiness stipulates economic advantage as its criterion for change.

Each phase of modernization has impacted most drastically on agriculturally marginal rural areas. Inability to adopt machinery or realize the gains from fertilization began the marginalization of many physically poorer farm areas, especially in eastern Canada. Given the concentration on fewer, larger production units (generally located on high capability land), the inelasticity of demand, and lack of any policies that specifically seek to maintain farming in any area, restructuring has led to the virtual demise of viable agriculture outside core areas and the progressive weakening of rural communities for which agriculture was at least part. Worst-hit areas have been Nova Scotia and New Brunswick, where less than 10% of farms remain, and in the peripheral areas of Quebec and Ontario where agriculture only survives in a few pockets.

The impacts have not stopped there. Rural farm communities in core zones have been impacted by both reductions in numbers of farms and farm families and by the propensity for industrial farming to bypass the local communities. The concentration of supply and processing facilities, the decline in farm labour demand, and the concentration on farm business *per se*, have reduced local employment in agricultural input, service and processing, and weakened socioeconomic and political life in rural communities. Hardest hit is the Prairie region, whose 'community system' has lost hundreds of its smaller service settlements. Communities that survive do so by diversifying away from agriculture. In southern Ontario, where there is an increasing separation of farm and non-farm rural systems, the latter are characterized more by retirement and recreation or as part of the rural-urban fringe. Decoupling represents several types of loss: by the farm community, many of whom are becoming isolated; by the rural sector, which is divorced from one of its definitial characteristics and risks losing its overall *raison d' être*; and by society at large for whom the rural system loses its coherence.

Summary and Conclusions

The foregoing material supports the premise that Canadian agriculture is undergoing a fundamental restructuring as part of the shift to a total 'agri-food' system based on the industrial model. Whereas Canadian agriculture developed on, and was for long dominated by the farm level, this century has seen its gradual domination by elements beyond the farm gate. Farmers were attracted to an industrial model by its technological inputs, economic efficiency and promise of financial success; a significant minority restructured their operations to meet industrial criteria, notably increased scale, capital intensity and specialization of production, together with strong linkages with all sectors of agribusiness. Governments have facilitated an integrated system, especially through marketing arrangements and policies aimed at output and income stabilization. Unfortunately, despite increased productivity per farm, per hectare and per person engaged in agriculture, Canadian farmers find themselves in a precarious financial position, dependent upon government subsidies to remain in business. Although this financial stress is, in part, outside national control, it calls into question the industrial model and the dependence it has engendered at the farm level. The problems of economic dependence on both the market and secondary inputs are underscored by the degrading effects of industrialized agriculture on rural physical environments, as well as the decoupling of farms and farm populations from rural communities.

While the conditions affecting Canadian agriculture have parallels in other developed market economies, other factors are exacerbating the situation. One is the decision to opt for free trade with the US, another involves negotiations under the GATT. The object of the Free Trade Agreement was to give access to US markets for Canadian goods, including some agricultural exports. This decision was made in relation to the overall economy, but it has had impacts on agriculture that were anticipated but not really appreciated (Troughton, 1991). The GATT negotiations include attempts to end the grain price war, which is destroying the western Canadian grain economy. These are laudable objectives, but free trade, both bilateral and international, poses a major threat to Canada's domestic market producers, especially those operating under supply management. While the federal government states that it has agricultural interests at heart, nevertheless, it seems willing to sacrifice agriculture within the larger context of free trade. The latest agricultural policy statement says, in effect, that agriculture will have to restructure further to meet the realities of global economic competition (Agriculture Canada, 1989).

At the present time, the hoped-for benefits of Canada-US Free Trade have not materialized, and the GATT negotiations seem to be on the verge of eliminating supply management, but not the grain subsidies. Consequently, while many farmers seem likely to continue to suffer from the effects of trade wars, those who have been protected, and who were the only group enjoying some form of security within the restructured framework, are now also threatened.

References

Agriculture Canada (1989) *Growing Together, a Vision for Canada's Agri-food Industry*. Publication 5268/E, Supply and Services, Ottawa.

Brinkman, G. (1988) *Structure of Canadian Agriculture*. Paper, Agricultural and Rural Restructuring Experts Meeting, Regina.

Buckley, H. and Tihanyi, E. (1967) *Canadian Policies for Rural Adjustment. A Study of the Economic Impact of ARDA, PFRA and MMRA*. Special Study 7, Economic Council of Canada, Ottawa.

Craddock, W.J. (1970) *Interregional Competition in Canadian Cereal Production*. Special Study 12, Economic Council of Canada, Ottawa.

Federal Task Force on Agriculture (1969) *Report: Canadian Agriculture in the Seventies*. Queens Printer, Ottawa.

Fuller, A.M. and Mage, J.A. (eds) (1976) *Part-Time Farming: Problem or Resource in Rural Development?* GeoAbstracts, Norwich.

Fulton, M. (1987) Canadian agricultural policy. *Canadian Journal of Agricultural Economics* 34, 107–25.

Fulton, M., Rosaasen, K. and Schmitz, A. (1989) *Canadian Agricultural Policy and Canadian Agriculture, a Study for the Economic Council of Canada*. Supply and Services, Ottawa.

Gilson, J.C. (1982) *Western Grain Transportation, Report on Consultations and Recommendations*. Supply and Services, Ottawa.

Hall, Justice E. (1977) *Grain and Rail in Western Canada. The Report of the Grain Handling and Transportation Commission*. Supply and Services, Ottawa.

Hennessey, S.G. (1965) *Report of the Ontario Milk Industry Enquiry Committee*. Ontario Department of Agriculture, Toronto.

Hill, S.B. (1985) Redesigning the food system for sustainabilty. *Alternatives* 12 (3/4), 32–6.

Kneen, B. (1989) *From Hand to Mouth: Understanding the Food System*. NC Press, Toronto.

Mitchell, D. (1975) *The Politics of Food*. Lorimer, Toronto.

Schnaiberg, A. (1975) Social synthesis of the societal-environmental dialectic: the role of distributional impacts. *Social Science Quarterly* 56, 5–20.

Special Committee on Farm Income, Ontario (1969) *Report: The Challenge of Abundance*. Ontario Ministry of Agriculture and Food, Toronto.

Troughton, M.J. (1986) Farming systems in the modern world. In: Pacione, M. (ed), *Progress in Agricultural Geography*. Croom Helm, London, pp. 93–123.

Troughton, M.J. (1989) The role of marketing boards in the industrialization of the Canadian agricultural system. *Journal of Rural Studies* 5, 367–83.

Troughton, M.J. (1991) Canadian agriculture and the Canada–US Free Trade Agreement: a critical appraisal. *Progress in Rural Policy and Planning* 1, 176–96.

Warnock, J. (1978) *Profit Hungry: The Food Industry in Canada*. New Star Books, Vancouver.

3

Sectoral Adjustments in Agriculture: Dairy and Beef Livestock Industries in Canada

Quentin Chiotti

According to the Canada–United States Trade Agreement (CUSTA), the 1990s was to be a decade of stability and relative prosperity for the Canadian agri-food system. In consideration of recent trade disputes, however, which reflect both domestic and international pressures for change in the institutional and economic environment within which this system operates, the Canadian agri-food system may, in fact, undergo considerable restructuring by the end of this century. By focusing on the changing institutional and economic environment, this chapter attempts to examine restructuring critically in two agri-food sectors: the dairy and beef livestock industries in Canada. Restructuring, in this context, is examined at a macro-scale level of analysis, and refers to the broader spatial and structural adjustments involving producers and processors within each industry.

From the outset, it is important to note that evidence of adjustment in each agri-food sector is, in part, a reflection of a diverse and complex set of forces that are reshaping the Canadian rural system. While these forces, such as the creation of more integrated continental markets and corporate restructuring by transnational agribusiness firms, clearly have international dimensions, they are also having a significant impact on Canadian agriculture. In consideration of these broader forces and their implications on the rural system, the primary objective of this discussion is more modest and more focused, i.e. to provide an analysis of the potential and actual restructuring in the Canadian dairy and beef livestock sectors, as a response to the changing institutional and economic environment. Although the discussion does not directly address the implications of agri-food restructuring on the Canadian rural system, it is hoped that this synthesis will nonetheless provide a foundation for further critical discourse in this area. To achieve this objective, the discussion is organized into three sections. In section one, the institutional and spatial structures for each industry are identified. Contemporary disputes are addressed in section two, highlighting the specific details concerning the recent institutional debates and linking these to the potential and actual restructuring response. Lastly, the discussion briefly outlines the broader spatial and structural dimensions of future adjustment in each industry that may occur as the Canadian agri-food system becomes further integrated within the North American market.

Institutional and Spatial Structure

In many respects, the dairy and beef livestock industries in Canada have much in common. In economic terms, they have historically been the two most significant sectors in the Canadian agri-food system, and since World War II they have both undergone considerable restructuring, involving processes of agricultural industrialization at the farm level and consolidation in the processing sector (see Chapters 1 and 2; also Mitchell, 1975; Warnock, 1978; Sundstrom, 1984). Despite these similarities, however, there are significant differences in the institutional arrangements governing, and the geographical distribution of, production and processing activity. A basic understanding of these differences is necessary in order to appreciate the broader causes and implications of restructuring within each agri-food sector.

The dairy industry is primarily oriented towards the domestic market and is closely regulated at both the provincial and federal levels of government. The production, processing and distribution of fluid milk falls under provincial jurisdiction, whereby marketing boards balance supply and demand for milk via production quotas (supply management). Milk produced for manufacturing purposes is regulated in conjunction with the federal government through the Canadian Dairy Commission and a national milk marketing plan that allocates production quotas to participating provinces. Under supply management, dairy producers are guaranteed a fair price for their milk, based on a cost-of-production formula, which serves to provide them with higher than average farm incomes, yet also gives rise to higher consumer prices. Restrictions on cheaper foreign imports are therefore critical to the successful management of this system. In contrast, the beef livestock industry is essentially a free enterprise system that is well integrated within a continental, if not global, market. State intervention has seldom ventured beyond the establishment and enforcement of health standards, although in recent years the state has intervened to regulate imports (via the Meat Import Act, 1981) and to provide modest subsidies to cattle producers (via the National Tripartite Stabilization Plan, 1987). Historically, the industry has enjoyed the relatively unimpeded cross-border movement of cattle and carcass beef between the US and Canada, and has also regularly drawn upon imports of manufactured beef from Australia and New Zealand.

Table 3.1. Farms with sales of C$2500 or more, classified as dairy or cattle, by province, 1986.

Province	Number of farms	
	Cattle	Dairy
Newfoundland	45	68
Prince Edward Island	652	584
Nova Scotia	979	698
New Brunswick	739	631
Quebec	5763	15 906
Ontario	17 160	11 028
Manitoba	4682	1412
Saskatchewan	7866	881
Alberta	17 110	1828
British Columbia	4266	1150
Canada	59 262	34 186

Source: Statistics Canada (1987).

Operating within different institutional arrangements, production and processing activities for each industry have somewhat different regional patterns (Table 3.1). Whereas the perishability of fluid milk has maintained a close spatial relationship between production and processing, the national milk marketing plan has institutionalized the number of dairy producers and manufacturing plants at historical levels of production; consequently, almost 80% of Canada's dairy manufacturing activity is concentrated in Ontario and Quebec. Despite this concentration, dairy policy has also functioned to preserve an industry in areas of 'comparative disadvantage'. In this sense, supply management is analogous to a regional development programme, and has susbsequently fostered a dairy industry that may be domestically viable, yet internationally uncompetitive. In contrast, most cattle slaughtering activity is concentrated in Alberta and Ontario, which have historically shared about 70% of the national beef market. A key feature of the beef livestock industry is that central Canada is essentially in a deficit position, although Ontario is technically self-sufficient in slaughter cattle. This separation of production and consumption has fostered an industry that has relied upon the massive movement from western Canada of carcass beef to Quebec and feeder cattle to feedlots in Ontario. Combined with traditional cross-border trade of cattle and beef, the end result is a domestic beef livestock industry that is complex and dynamic in its spatial and structural organization, and also marginally competitive within a North American market.

Contemporary Disputes and Trends

When the Canada-US Free Trade Agreement was implemented on 1 January 1989, the Government of Canada was assuring the agricultural community that the 1990s would be a decade of prosperity for Canadian farmers (External Affairs, Canada, 1988). This view could also be extended beyond the farmgate, as it was anticipated that meat packers would have secured access to US markets, while dairy processors (and supply management) would continue to be protected from foreign competition. Under this scenario it was unlikely that the Canadian agri-food system of the 1990s would undergo a significant reorientation away from the traditional patterns of agricultural production and food processing that existed during the previous two decades. However, entering the third year under CUSTA, it is generally recognized by numerous producers, agribusiness firms, agricultural economists and even government policy-makers that the purported benefits of the Agreement were naively optimistic and deserve critical re-evaluation.

Recent Canadian agricultural trade disputes, such as those involving the movement of beef, hogs and dairy products, tend to illustrate that some of the concessions and safeguards incorporated within CUSTA may, in fact, be more rhetoric than reality. Contradictions between CUSTA and the GATT, and the unresolved conflicts between them, have created an uncertain institutional environment that has producers and processors of meat and dairy products extremely worried about the future of their respective industries. The stalemate that had developed by January, 1992, in the Uruguay Round of GATT negotiations has exacerbated these problems, as a reduction in grain subsidies and the clarification of 'like product' were fundamentally important for the continuation of a secure and relatively productive Canadian agri-food system. Depending upon the resolution of these conflicts, there is a distinct possibility that agricultural production and processing in Canada, and the marketing relationships between them, will undergo a considerable transformation during the 1990s; this restructuring process has, in fact, already begun in some agri-food sectors as a response to institutional pressures and competitive market forces.

Institutional debates

The articles pertaining to agriculture in CUSTA are the most extensive in the Agreement, and in many cases are tied within the broader rules and regulations governed by the GATT. While the articles are presented as being consistent and non-conflicting, upon closer examination a number of contradictions between them and in relation to the multilateral trade negotiations can be identified. Many of these contradictions have already been addressed elsewhere in the literature (see Warnock, 1988; Chiotti, 1991; Troughton, 1991), so they will not be reviewed in depth; however, a few key contradictions deserve mention. The overall tone of CUSTA is clearly expressed in Chapter 1, and particularly Article 102, which includes in the objectives of the Agreement provisions to eliminate trade barriers and to facilitate conditions of fair competition within Canada and the US. Notwithstanding this commitment to freer trade, however, the CUSTA suffers from some fundamental contradictions, specifically in relation to the goals to reduce trade-distorting subsidies and to safeguard public welfare. This seemingly contradictory position, between working together towards the reduction of trade-distorting subsidies while somehow also maintaining a safety net of social programmes, is particularly illustrated in the chapter pertaining to agriculture (Chapter 7). Under Articles 701 (Agricultural Subsidies) and 703 (Market Access for Agriculture), Canada and the US are committed to working towards the elimination of all agricultural trade-distorting subsidies on a global basis, while improving access to each other's markets through the reduction of import barriers. With Article 710 (International Obligations), however, Canada also seems to have retained the right to protect supply management under GATT Article XI.

Combined with the provisions to restrict imports of poultry and eggs to historic levels (Article 706), Canadian dairy producers believed that supply managed commodities were well protected under CUSTA. In fact, the official government position, which was supported by agricultural economists and provincial policy-makers, initially assured milk producers that supply management was not on the bargaining table. Furthermore, it is clearly stated within the CUSTA document that 'Canadian dairy farmers will continue to benefit from supply management programs since these are not affected by the Agreement and are consistent with Canada's GATT obligations' (External Affairs Canada, 1988, pp. 76–7). These claims continued to be made by the Government and agricultural economists throughout the first year of CUSTA, despite the fact that supply management and import quota restrictions had been clearly identified as 'trade distorting' by the US Government (USDA, 1987) and the GATT community (Miner and Hathaway, 1988).

In contrast with the potential inconsistencies in the dairy provisions, improved access for beef processors (and indirectly for cattle producers) seemed to be relatively straightforward in the Agreement, with the meat provisions clearly oriented towards facilitating the movement of beef and pork across the Canada-US border. Article 704 (Market Access for Meat) provides for the mutual exemption of each country from restrictions under meat import laws, restrictions that have historically limited the amount of imported beef to specific quota levels. Under Article 708 (Technical regulations and standards for agricultural, food, beverage and certain related goods), both countries committed themselves to harmonizing technical regulations, thereby establishing an 'open border' policy for meat inspection. Given the relatively similar inspection systems in both countries, the intent of this provision was to reduce the frequency of border inspections for meat to levels that would occur for shipments between plants within a country. It was mutually agreed upon by both parties that a one in twenty frequency rate would be enforced for entry point inspections (von Massow *et al.*, 1991). Combined, these provisions would ensure the freer movement of beef and pork while securing access for Canadian producers and processors to the larger US consumer market.

Challenges to these provisions began a few weeks before the Agreement was implemented, when the Minister of Agriculture responded to the concerns of Canadian dairy farmers and took further action towards securing Canada's borders from cheaper US competition. Ice cream and yogurt were specifically added to the Import Control List, and since GATT Article XI (Section 2 ci) allows governments to protect domestic agricultural production by placing quantitative restrictions on the imports of 'like products', the Minister's actions seemed to fall within Canada's legal obligations under the GATT. Canada subsequently unilaterally established a quota for US imports of ice cream and yogurt that reflected historical levels of trade between 1985 and 1987. Responding to an application submitted by the Pillsbury Co. of Minneapolis, Minnesota, to export 'Haagen-Daz' ice cream into Canada (at the time domestically manufactured under license by Wm. Neilson, a subsidiary of the George Weston agribusiness empire), the United States Department of Agriculture (USDA) lodged a formal complaint through the GATT against Canada's import quotas. The US challenge was based upon the ambiguity surrounding the definition of a 'like product', plus the contention that Canada had breached an understanding amongst the GATT parties that they would refrain from additional trade restraints for the 4-year period of the Uruguay Round of negotiations (Gould, 1989; Warley, 1990).

The Canadian response to the USDA challenge was predictably negative, interpreting the action as a 'hostile' assault on the integrity of the supply management system in Canada. Indeed, the very survival and successful functioning of supply management is dependent upon the restriction of imports. Canada's defence was based upon two central arguments: (i) that quantitative import restrictions are legal under the GATT, and (ii) under supply management Canadian milk producers do not produce surpluses (and thereby do not contribute to price instability on the world market). Furthermore, the USDA actions have been questioned as 'hypocritical', since the US is able to protect its own domestic dairy industry under a relatively unpublicized 1955 GATT waiver. Made permanent in 1970, the waiver allows the US to implement its own import quotas on ice cream, which in 1987 was set at 400 000 gallons per year, or less than 0.05% of their total domestic production.

A GATT panel ruled, in September 1989, that Canada had violated its obligations under international trading agreements, deciding that ice cream and yogurt were not 'like products' with raw milk, since their manufacture required further processing than simply fluid milk. Consequently, Canada could not limit such imports under the premise that they harmed domestic dairy producers. Canada initially challenged this decision on technical grounds, and then shifted its efforts by tabling a proposal for strengthening and clarifying GATT Article XI in support of supply management programmes. Canada's position, which has been supported during most of the Uruguay Round by Japan and the European Community (EC), runs counter to the most recent GATT proposals (January, 1992) that favour the conversion of import quotas into tariffs.

While these and other efforts to promote the virtues of GATT Article XI illustrates that the Government of Canada has not abandoned its public support of supply management, there is growing evidence to suggest that its true position is definitely vacillating, if not shifting towards what one agricultural economist (Warley, 1990) has described as processor survival and sectoral development of the Canadian agri-food system. In 1989, for example, there were two government reports (de Grandpré, 1989; Agriculture Canada, 1989) that focused upon processor survival and profitability, while stressing the need for the supply management system to be more 'market responsive'. A recent federal task force on National Dairy Policy (Agriculture Canada, 1991a) has reiterated this philosophy, arguing that milk producers and processors must undergo considerable restructuring (to achieve greater economies of scale and labour efficiency) and rely upon less government subsidies, in order to be competitive in a freer world marketing trading environment. The federal government's position also tends to reflect increasing internal pressure to alter the structure of supply management in Canada. In this case, it seems that consumer organizations, economists and processors have used the controversy as an opportunity to voice considerable dissatisfaction with the marketing board system, which tends to increase the price of raw agricultural inputs to processors (and eventually consumers), to levels above those that could be obtained under a free market system. Ault's dairies, for example, which is the largest dairy in Canada and is purported to control over 50% of the Ontario fluid milk market, has already publicly threatened to close down production, unless raw milk prices are lowered to US levels (Canadian Press, 1990). In response to increasing domestic and international pressure, producers have been taking an increasingly active role to educate the Canadian public, policy-makers and even members of the GATT community, with regard to the benefits of supply management (Dairy Farmers of Canada, 1991).

In contrast, two disputes in the beef livestock industry involve entirely different issues and participants. The Canada-EC dispute over imports of Irish beef basically involves much more than merely the importation of 'manufacturing beef', but also reflects Canada's commitment towards an open trading relationship with the US (Kerr and McGivern, 1990; van Duren and Martin, 1990). Essentially, a dramatic increase in Irish exports between 1981 and 1984 prompted the Canadian Cattlemen's Association (CCA) to pressure the federal government to invoke the Meat Import Act in 1985. This action was followed by the initiation of a countervail suit under the Special Import Measures Act, which was referred to the Canadian Import Tribunal (CIT) when the claim was contested by the EC, the Irish Livestock and Meat Board, and one Canadian importer. The case was fought primarily on economic arguments, and although material injury had not been sustained, the Tribunal ruled that material injury would occur if imports were re-established at previous levels. It was argued that the substitution of domestic production for Irish beef would result in Canadian cows flooding the US market, which would then result in US counteraction to restrict entry of Canadian beef and cattle. In October 1986, a GATT panel was established to hear an appeal to the Tribunal's decision, and ruled against Canada's restrictions. Basing their decision upon legal grounds, the CCA was ruled as not having 'standing' in bringing the original suit before the CIT, since farmers produce cattle and not beef (essentially ruling that beef and cattle were not 'like products'). Fortunately for Canadian cattle producers, the acceptance of the GATT decision requires a consensus, and since the US, Australia and New Zealand have yet to accept the Panel's report, Canada has continued to restrict the importation of Irish beef. In fact, Canada's countervail measures have recently been renewed by the CIT for an additional 5 years (Brinkley, 1991).

Ironically, Canada's commitment to the North American market, as secured under CUSTA, has recently been undermined by the meat inspection dispute with the US. This dispute differs from the others such that it involves an issue of non-tariff barriers to trade. Furthermore, since the meat inspection issue was to represent the model for the harmonization of technical standards, US actions subsequently contravene the fundamental spirit of the Agreement (von Massow *et al.*, 1991). Essentially, the dispute involves an increase in the number of cross-border inspections of Canadian beef by US meat inspectors, which is directly contrary to the intent of the provisions in CUSTA. The problem actually began shortly after the acceptance of CUSTA, when the US Department of Agriculture's Office of the Inspector General recommended that point of entry checks be made to ensure the wholesomeness of meat imports. In 1988, the USDA partially privatized their border inspection service in an effort to reduce the costs of upgrading, whereby government officials would conduct the inspections, but the facilities would be owned by private individuals or companies. Since CUSTA was to reduce the number of meat inspections, the private owners of these facilities feared that their investments would eventually become obsolete. Consequently, during the past 2 years the inspections have become more frequent and more costly, with inspections typically rising from one in twenty to one in seven loads and the cost increasing from C$200 to over C$700 for each individual truck inspected. Obviously, the cost of shipping beef to the US is becoming prohibitively expensive under these conditions. Despite a February 1990 agreement between Canada's agricultural minister and the US agriculture secretary to end re-inspection of meat products crossing the border, the 1-year experiment has yet to be implemented and this trade dispute remains unresolved. The ultimate impact of this dispute will extend throughout the Canadian beef livestock industry, beginning at the processing level and then filtering down to cattle producers. According to von Massow *et al.* (1991), with access not assured to the US market, there will be limited investment in Canadian processing plants.

The restructuring response

If the Canadian dairy industry (and other supply-managed industries) have to compete on a 'level playing field' with their US counterparts, then considerable restructuring is likely to occur at both the farm and processing levels of production. Basically, Canadian dairy policy has nurtured the development of an internally viable industry, but there is considerable uncertainty concerning its continental competitiveness. While supply management has undoubtedly facilitated the industrialization of agriculture (Troughton, 1989), Canadian dairy farms are generally smaller and less efficient than their American counterparts. Marketing boards have tended to maintain the family farm as the dominant mode of production; consequently, there are only a handful of 'mega farm agribusinesses' in Canada, while such operations are more common in the Pacific and Southwestern regions of the US (Agriculture Canada, 1991b). Meanwhile, in the processing sector, Canadian firms are also generally smaller and less efficient than their US competitors (OMAF, 1988a, b), particularly in terms of manufactured milk products. Higher input costs (industrial milk), improvements in transportation, and in some cases identical brand names, combine to make the Canadian dairy processing sector quite vulnerable to US competition. Quite simply, if Pillsbury can supply the entire US market with premium ice cream from only two plants (in New Jersey and California), it seems obvious that the large processors south of the border could easily expand their productive capacity to meet consumer demand in Canada.

In consideration of these structural differences, there is great potential for radical restructuring in the Canadian dairy industry, especially in marginal areas of production, such as Northern Ontario and the Atlantic provinces (Chiotti, 1991). The emphasis here, however, lies more in potential than actual restructuring, for the only tangible evidence of rationalization since the implementation of the Agreement has occurred among the major dairies engaged in the processing of fluid and industrial milk products. For instance, recent purchases by John Labatt Ltd of the ice cream division of Neilson's dairies and the cheese division of Canada Packer's has solidified the position of Ault's Dairies as the dominant milk processor in Ontario (Bertin, 1991). Both acquisitions were facilitated by corporate restructuring in the parent companies of the acquired firms, as a response to the changing institutional environment and changing competitive forces. Illustrative of the market value of brand names, and the mobility of industrial capital, the realignment of corporate ownership has resulted in some job losses; however, this corporate restructuring may have also created stronger Canadian agribusiness firms engaged in dairy, meat-processing and confectionery activities, enabling them to compete more effectively in a North American market. Labatt's dairy operation, for instance, which has secured access to cheaper American supplies through a strong processing and distribution network in northeastern US, is well placed to compete successfully with any dairy in Canada (including many of their larger US counterparts) within a freer market environment.

With respect to the Canadian beef livestock industry, in recent years this sector has already undergone considerable restructuring, partly in anticipation of greater integration with the US agri-food system. The radical restructuring of the Canadian beef livestock industry can best be described as a westward shift from Ontario to Alberta, although there is growing evidence to suggest that this outward movement is occurring in all other provinces. Reflecting a process of decentralization from major urban centres in central Canada to small towns in western Canada, this western shift has significantly altered the traditional patterns of production, processing and interprovincial trade. As shown in Table 3.2, the beef-packing industry in Ontario has dramatically declined, whereas Alberta's share of the national market is now over 50%. A number of factors have facilitated this movement, including differentials in provincial subsidies to cattle producers and processors, and the activities of multinational agribusiness corporations in the beef-packing industry; however, the most significant factor is the low feed grain prices that have resulted from the yet to be resolved global grain subsidy war between the US and the EC.

The resulting economically favourable environment in Alberta for beef production has been further enhanced by the 1988 arrival of Cargill Inc. into the Canadian beef-packing industry (Kneen, 1990). Cargill Inc., the largest privately owned company in the US, with extensive American

Table 3.2. Number of cattle slaughtered in federally inspected establishments in Ontario, Alberta and Canada, 1981–90.

Year	Ontario		Alberta		Canada
	Number	%	Number	%	Number
1990	548 599	21.2	1 312 654	50.7	2 587 239
1989	693 462	24.6	1 271 321	45.1	2 818 936
1988	730 342	26.3	1 209 997	43.6	2 773 638
1987	802 751	27.9	1 135 084	39.4[a]	2 879 455
1986	872 795	28.0	1 126 256	36.1[a]	3 118 401
1985	839 729	26.6	1 270 408	40.2	3 159 307
1984	827 261	26.5	1 261 027	40.5	3 116 220
1983	951 773	29.4	1 308 027	40.4	3 241 682
1982	983 526	29.9	1 320 551	40.1	3 293 947
1981	1 031 419	32.3	1 294 827	40.5	3 196 887

[a] Relative decline due to labour difficulties.
Source: Agriculture Canada, *Livestock Market Review (1981–1990).*

and global operations in feedlots and meat-packing activity, has recently constructed a C$53 million 'state of the art' meat-packing plant with the assistance of a C$4 million grant from the government of Alberta. Reputed by Cargill to be the most efficient beef plant in North America, the fully integrated beef facility has a single shift capacity of 6000 head per week, and is essentially gearing production for the Canadian, California and Pacific Rim markets. Located within a 250 km radius of 80% of the fat cattle produced in Alberta, the High River plant is strategically located to take over the industry. Since its opening in 1989, Cargill Inc. has been trying to buy its way into the market by undercutting the price of beef and hiring non-unionized labour, at a wage rate C$2–C$3 an hour lower than its major competitors.

The initial consequence of Cargill's strategy has been the retention of Alberta cattle within the province, and the attraction of feeder cattle and calves to Alberta feedlots, in response to the higher prices. On a regional basis, the effect has been to further concentrate feedlot operations in Alberta, while subsequently drastically reducing the number of stockers (feeder cattle and calves) that move westward to Ontario feedlots. This movement of feeder cattle has declined from almost 700 000 in 1980, to less than 200 000 in 1990. With less cattle to draw upon, and the high costs associated with the transportation of slaughter cattle, the entire Ontario livestock industry soon found themselves in a severe state of crisis by the late 1980s (Grier, 1988; McDermid, 1988). The situation has become so acute that the largest and most efficient beef-packing plant in Ontario has been forced to import carcass beef from the US to maintain its full capacity. Combined with outdated plants and an increasing overcapacity, the industry has undergone a substantial downsizing in recent years. In the past 3 years, for example, a total of 10 plants have closed or shifted production into hogs or further processing (Table 3.3), resulting in an estimated loss of 12 000 head per week in slaughter capacity. The most notable exit from the industry has been Canada Packers Ltd (as part of Hillsdown's restructuring plan), who converted their Ontario operations to either hog slaughtering or distribution, and have just recently divested themselves of their four western packing plants.

Table 3.3. Status of federally-inspected plants in Ontario slaughtering cattle, 1988–90.

Plant and location	Status
Beef Terminal, Toronto	Active
Better Beef, Guelph	Expanded
Bolton, Bolton	Closed 1988
Bruce Packers, Paisley	Active
Canadian Dressed Meats, Toronto	Closed 1989
Comfort and Tylee, St Ann's	Active
Dorr Brothers, Stoney Point	Expanded
Fearman, Burlington	Shift to hogs, spring 1990
Grace, Toronto	Closed 1990
Gross, Toronto	Active
Hunnisett, Toronto	Closed
Lefaivre, Lafaivre	Closed 1990
MGI, Kitchener	Active
Northern Beef, Proton Station	Closed 1988
Peterborough Packers, Indian River	Closed 1989
Schneider, Kitchener	Shift to processing, 1989
St Helen's, Toronto	Active
Star Brand, Binbrook	Active
Tenderlean (Canada Packers), Burlington	Shift to distribution, 1990

Source: Grier (1988) and compilations by the author.

Whereas Ontario has already experienced the considerable downsizing of its beef-packing industry, the situation in Alberta is unclear. A major problem is that the government of Alberta seems to be determined to support the meat-packing industry, regardless of the costs. Substantive financial support has been provided to other major agribusiness firms, such as loans and grants to Centennial Packers and Lakeside Packers (partially owned by Mitsubishi Canada Ltd), including the complete provincial takeover of the Gainer's operations in 1989. If Cargill Inc. installs a second shift in production, however, the productive capacity of Alberta will exceed 40 000 head per week, which is approximately 15 000 head in excess of their 1989 average kill. In consideration of this overcapacity and restricted access to the US market, there is little doubt that the bottom will eventually fall out of the Alberta beef-packing industry, due to either chronically slim profit margins or when the provincial government encounters a predicted fiscal crisis. It is estimated that only three large plants will remain in western Canada, specifically two 'mega plants' (Cargill and Lakeside) and one medium-size plant (with a capacity of 5000 head).

Future Scenarios

There can be little doubt that the dairy and beef livestock industries in Canada will undergo further restructuring during the 1990s. The extent and spatial pattern of this restructuring, however, is contingent upon a variety of economic and political factors, many of which are difficult to predict. Nonetheless, a safe and broad prediction concerning restructuring at the macro-scale involves each industry becoming increasingly market-responsive to consumer demands and competitive pressures, as Canadian producers and processors become further integrated within a North American agri-food system.

Assuming that the meat-inspection issue and the right to restrict beef imports from the EC are resolved in Canada's favour, the Canadian beef-livestock industry will continue along its recent restructuring path. This is likely to involve the further breakdown of interprovincial trading relationships between western and central Canada, and the integration of these regions within the US market. While Alberta will continue to attract feeder cattle from other provinces, the beef-processing sector will undergo considerable rationalization. The extent and form of restructuring in the dairy industry is largely contingent upon the fate of GATT Article XI, and Canada's ability to maintain its supply management system. If supply management is dismantled, one can expect significant rationalization in the processing sector, and the further concentration of milk production towards regions of comparative advantage (primarily in southern Ontario and Quebec).

What is more difficult to predict, however, concerns the impact of restructuring at a regional or micro-scale; particularly in terms of the impact upon the rural system. In an effort to become globally competitive, any attempt to achieve greater efficiency and economies of scale will also result in the further spatial concentration of production and processing activity, and a corresponding decline in marginal areas of production. Given the potential for job loss and a reduction in the number of viable farms, one needs to consider seriously the impact of these changes, particularly in residual production areas. For agricultural policy makers who predicted a decade of prosperity for the Canadian agri-food industry, their research efforts should be directed towards these issues.

References

Agriculture Canada (1989) *Growing Together: A Vision for Canada's Agri-food Industry*. Publication 5269/E, Ottawa.

Agriculture Canada (1991a) *Growing Together: Evolution of the Canadian Dairy Industry*. Report of the Task Force on National Dairy Policy, Ottawa, 31 May.

Agriculture Canada (1991b) *Growing Together: A Comparison of the Canadian and US Dairy Industries*. Report of the Task Force on National Dairy Policy, Appendix E, Ottawa, February.

Agriculture Canada *Livestock Market Review*. 1981–1990, Supply and Services, Ottawa.

Bertin, O. (1991) Labatt raises stake in dairies. *Globe and Mail* 8 January, Toronto.

Brinkley, A. (1991) Ruling has cattle farmers breathing easier. *Daily Mercury* 24 July, Guelph.

Canadian Press (1990) Milk prices may force closing, Ault says. *Globe and Mail* 21 March, Toronto.

Chiotti, Q. (1991) Farmers in a state of purgatory: the Canadian dairy industry in the aftermath of the Canada–US Free Trade Agreement. In: Beesley, K. (ed.), *Rural and Urban Fringe Studies in Canada*. Geographical Monograph 21, Atkinson College, York University, Toronto, pp. 239–67.

Dairy Farmers of Canada (1991) *Supply Management Perspectives of the GATT Negotiations*. Mimeo, Ottawa.

de Grandpré, A.J. (1989) *Adjusting to Win*. Report of the Advisory Council on Adjustment, Ottawa.

External Affairs, Canada (1988) *Trade: Securing Canada's Future*. (The Canada–US Free Trade Agreement) Government of Canada, Ottawa.

Gould, P. (1989) High stakes game at the GATT table. *Ontario Milk Producer* June, 29–31.

Grier, K.J. (1988) *Ontario Beef Packer Situation Outlook*. Ontario Ministry of Agriculture and Food, Toronto.

Kerr, W.A. and McGivern, D.B. (1990) The dispute concerning Canadian imports of Irish beef: a problem of multilateral production substitution. In: Lermer, G. and Klein, K. (eds), *Canadian Agricultural Trade: Disputes, Actions and Prospects*. The University of Calgary Press, Calgary, pp. 93–106.

Kneen, B. (1990) *Trading Up*. NC Press, Toronto.

McDermid, K. (1988) *Beef Marketing Task Force Report*. Ontario Ministry of Agriculture and Food, Toronto.

Miner, W.M. and Hathaway, D.E. (1988) World agriculture in crisis: reforming government policies. In: Miner, W.M. and Hathaway, D.E. (eds), *World Agricultural Trade: Building a Consensus*. The Institute for Research on Public Policy, Halifax, pp. 37–110.

Mitchell, D. (1975) *The Politics of Food*. Lorimer, Toronto.

Ontario Ministry of Agriculture and Food (1988a) *Competitiveness of Selected Industries in Ontario's Food Processing Sector Under the Canada–US Free Trade Agreement*. Woods Gordon, Toronto.

Ontario Ministry of Agriculture and Food (1988b) *Assessment of the Impacts of the Canada–US Free Trade Agreement on the Ontario Agriculture and Food Sector*. Ontario Ministry of Agriculture and Food, Toronto.

Statistics Canada (1987) *Census of Canada, 1986: Agriculture, Canada*. Supply and Services, Ottawa.

Sundstrom, M. (1984) The Canadian dairy industry in transition: 1950–1980. In: Bunce, M.F. and Troughton, M.J. (eds), *The Pressures of Change in Rural Canada*. Geographical Monograph 14, Atkinson College, York University, Toronto, pp. 34–53.

Troughton, M.J. (1989) The role of marketing boards in the industrialization of the Canadian agricultural system. *Journal of Rural Studies* 5, 367–83.

Troughton, M.J. (1991) An ill-considered pact: the Canada–US Free Trade Agreement and the agricultural geography of North America. *Trade Liberalization and Rural Restructuring in Canada*. Working Paper No. 1, The Rural Development Institute, Brandon.

United States Department of Agriculture (USDA) (1987) *Government Intervention in Agriculture*. ERS. Staff Report, No. 229, (April), USDA, Washington DC.

van Duren, E. and Martin, L. (1990) The impact of subsidized imports of low-quality beef from the EC on the Canadian cattle industry. In: Lermer, G. and Klein, K. (eds), *Canadian Agricultural Trade: Disputes, Actions and Prospects*. University of Calgary Press, Calgary, pp. 107–24.

von Massow, M., van Duren, E. and Martin, L. (1991) *Resolving Trade Disputes with the United States: A Level Playing Field or a Vehicle for US Vested Interests*. Discussion Paper DP91/01, The George Morris Centre, University of Guelph, Guelph.

Warley, T.K. (1990) *International Pressures on Supply Management*. Unpublished paper presented to the OAC Outlook Conference, University of Guelph, Guelph, 2 January.

Warnock, J.W. (1978) *Profit Hungry: The Food Industry in Canada*. New Star Books, Vancouver.

Warnock, J.W. (1988) *Free Trade and the New Right Agenda*. New Star Books, Vancouver.

II

Diversification of the Farm Business

4

Uneven Agrarian Development and the Social Relations of Farm Households

Richard Munton, Terry Marsden and Neil Ward

In previous papers we developed a typology of farm businesses. In these papers we argued that the parameters of the typology must extend beyond 'the business itself, to the less direct relations of production, involving external influences affecting production relations on the farm' (Whatmore *et al.*, 1987a, p.27; 1987b). Our methodology led, first, to the construction of a set of theoretically informed 'ideal types' by which farm businesses were classified according to their internal and external production relations and the degree to which they had been subsumed by off-farm capitals. A second stage sought to refine and develop the analysis by exploring a range of contingent conditions using empirical evidence from three lowland (west Dorset, east Bedfordshire and the Metropolitan Green Belt) and two upland (west Cumbria and north Staffordshire) farming areas in England. Over 400 farms were included in the study, surveyed between 1985 and 1988.

The field evidence revealed that external capitals penetrated farm production relations more extensively through mechanisms of indirect or 'formal' subsumption, such as technological, credit and marketing relations, than through direct or 'real' subsumption under which the ownership and management of the means of production are transferred from the farm family to the external capital. This finding requires that the relations between external capitals and internal farm production processes be viewed as in a constant state of interaction. The nature of the interaction is influenced not only by wider developments within the economic system but also by changes in the specific circumstances of family businesses, such as marriage, succession, illness, which, in combination, affect the pattern and degree of capital penetration. The evidence also undermines those theoretical positions that suggest that the processes of family farm development are either uniform or unilinear. The analysis confirms that although broad tendencies can be identified in the processes of subsumption, they are regularly contradicted at the individual farm level. These variations are accounted for, in part, by the differing options available to farm families in different localities (Munton *et al.*, 1988; Marsden *et al.*, 1989). In spite of these qualifications, the typology still provides a set of ideal types and hence a series of consistent arguments through which the diversity of empirical cases can be explored. The methodology employed is described in full elsewhere (Whatmore *et al.*, 1987a, b) and is not repeated here.

This chapter represents an extension of our earlier work. It attempts to incorporate directly aspects of the social reproduction of farm families into the analysis of subsumption. Furthermore, we recognize that in the 1990s the very nature and position of agricultural production in advanced economies, and the social relations upon which they are based, are subject to increased instability and debate as to their future.

Integrating Social Reproduction as an Internal Dynamic

Context

We wish to respond to Friedmann's critique that 'the division of labour, patterns of domination and struggle, the cyclical life of the enterprise' and the ways these are 'shaped by gender and generation' (Friedmann, 1986) are no more than alluded to in the arguments of agrarian political economists. Their arguments fail to capture, she suggests, the diversity of capital–labour relations on family farms (see also Whatmore, 1991). We agree with these arguments, although we cannot address more than a small part of her empirical agenda in this chapter. Specifically, we wish to examine the kinds of relations that are present at the point of production between capital's need to accumulate and the family's wish to maintain its control from generation to generation over the ownership and management of the farm business. This enquiry also raises broader questions. These include how such relations might be affected by state policies towards agriculture, expectations in society about the distribution of assets between family members, and the general profitability of farming (see Marsden and Murdoch, 1990).

The changing position of family farming under capitalism has generated an enormous literature. Here we are only concerned with the ability of farm families to ensure succession within the family and the level of commitment that families have to that goal. All the major reviews of family farming give particular prominence to this question but caution against making generalizations on the basis of evidence that is time and place specific. For that reason the following remarks draw very largely upon evidence collected in the United Kingdom since the 1950s. By historical standards, this period has been one of relative prosperity for farming achieved through the state management of markets and the promotion of an industrial model of production. The post-war period has been marked by a set of policy objectives that have placed considerable demands on the supply of capital and growth in the size of farm business. Until the early 1980s, agricultural policy was very largely directed towards increasing productivity, with gains in efficiency realized through technological innovation and scale economies (for a discussion of these issues see Bowers, 1985; Hill, 1990; Munton *et al.*, 1990).

As economic units, farm businesses have grown substantially in size over the period. At the same time, changes in social attitudes have undermined traditional patriarchical family relations leading to more complex distributions of resources and management control among family members. Most farm businesses, and the land they occupy, are no longer held in single ownership but in family partnerships or even more complex arrangements with non-kin representatives of off-farm interests (Whatmore *et al.*, 1990). The immediate reason for these changes may lie mainly in tax avoidance strategies and attempts to realize non-agricultural property rights but broader social tendencies favouring the more equitable treatment of spouses, including divorce settlements, and a more respected role and status for women on farms, have all contributed to the changing pattern. These trends may also create difficulties at the moment of succession if the inheritor has to buy out the interests of other family members. Another related social change is the reduced willingness among farm families to invest in the business, revealing the higher priority they now give to personal consumption (Reid, 1981).

Research into family farming in Britain would, on the whole, confirm the view that the primary management objectives of farmers are to maintain family control and to pass on an economically secure business to the next generation. These objectives demand a strategic planning horizon that is intergenerational and may, therefore, conflict with the interests of capital. In so far as this is so, then any increase in the penetration of farm businesses by external capital may require the family to enter into a range of compromises between its objectives and those of capital. Some evidence, drawn from the Mid-West of the United States, suggests that communities of 'yeoman' farming families that have sought a prudent relationship with external capitals have continued to flourish over long periods, while those consisting of 'entrepreneurial' farm businesses have not fared so well because of the latter's susceptibility to financial collapse in periods of economic stress (Salamon, 1985; Rheinhardt and Barlett, 1989). These authors also acknowledge, in agreement with Mooney (1988, p. 217), that, for the argument to hold, families must have 'access to strategic land, capital, water and other resources ("privileges") to survive'.

We would take this argument further, suggesting that where owner-occupation is the dominant tenure, it is not access (succession) to the land as such that is crucial but the cost of that access to the occupier. Those facing penal inheritance taxes, or those who have to buy out the interest of other family members, may inherit a non-viable business; and this possibility is made more likely where land prices are higher than can be justified from their agricultural returns for reasons of non-farming demands on the land. These considerations draw attention to a related issue, the changing need for capital over the family life cycle (Bennett, 1982; Moran, 1988), and the question of timing (Marsden *et al.*, 1987). The needs of capital and the family have to be synchronized in order to allow capital to be a liberating force. However, unlike previous periods, the present is characterized by high and variable interest rates and rapid but uneven technological change, which together complicate the family experience (Hawkins, 1991). It is no longer possible to view the generational period (say 30 years) as a single cycle (see, for example, Nalson, 1968), during which periods of high capital need (cost) can be compensated for by periods of low need. Rather, the agricultural treadmill is continuous, with fluctuations within it often depending on macroeconomic considerations. To invest at the 'wrong' moment, in terms of swings within the macroeconomy, may lead to consequences that cannot be rectified without progressive loss of family control, however apposite the timing seemed from the family 'stage' point of view.

But what is meant by continuity? It is necessary to distinguish between continuity in farming and continuity of occupancy of the same piece of land or farm. Evidence for Britain confirms a high level of hereditary association with the industry. According to Gasson *et al.* (1988), almost all surveys indicate that between 60% and 80% of farmers had farming fathers, but less than half inherited their parents' farms. The hereditary link is, however, thought to be increasing because of the growth in owner-occupation and the limited number of new tenancies awarded to 'new blood' occupiers. These trends have made the capital cost of entry prohibitive to all outsiders bar the most affluent or those seeking part-time smallholdings (Northfield Committee, 1979; Crow, 1987). Furthermore, to date there have been no surveys that record a substantial lowering of the expectation that most farmers have of one or more of their children succeeding to the business, and it is rare for farmers to identify and plan for a non-family successor (Harrison, 1975). It is at this point that the distinction needs to be made between inheritance (transfer of legal ownership and possession) and succession (the gradual transfer of management control from one generation to another, see Hutson, 1987). These two processes may occur in tandem and both have tended to come earlier in the life cycle in recent years in response to social trends (people retiring earlier), fiscal pressures (the need to plan ahead to minimize tax liability), and rate of technological advance.

The financing of the processes can create difficulties, notably the potential conflict between farm business efficiency and ensuring the equitable treatment of all children. Furthermore, it is reasonable to anticipate that with much of farming in recession and ever-widening opportunities for off-farm employment, the allure of managing the family business must be in decline. The current growth in pluriactivity (Shucksmith *et al.*, 1989; Fuller *et al.*, 1990) is not only reducing the significance of the farm as the source of family income but also diminishing its social role in defining the behaviour of its members. In addition, while the economic pressures on the family may be greater on small farms, the wider social networks of those from more affluent farm families will bring higher career aspirations and economic expectations. Furthermore promotion of the farm as a business – the dominant post-war model – may mean that family members' attitudes are now more determined by what they can 'get out' of the business than by what they think they need to 'put in'. In other words,

> the successful financial performance and continuity of the farm as a family business depend not only on the satisfactory functioning of the business enterprise but also on the continuity of family structures and processes
>
> (Gasson *et al.*, 1988, p. 34)

It is to highlight that what is meant and understood by the 'family' and the 'business' are in rapid flux, and a major reason for this is that what is meant by 'success' is increasingly being re-evaluated by attitudes established from outside the industry.

The survey evidence

The complexity of the issues raised in this brief review of the social reproduction of family farms cannot be easily captured empirically. Nonetheless, what follows is an initial attempt to incorporate the broad thrust of the argument over family continuity into the analysis of subsumption. To this end, farm principals were asked a number of questions relating to family structure and occupance, and what their expectations were for a next generation family member to succeed to the business. Just under 30% of farmers said that succession was definitely ruled out, while 53% saw succession as likely. The remaining respondents were unsure either because of the absence of children at the time of survey, the young age of their children, or because they were uncertain whether succession was an economically viable option. An important qualification to these findings is that the respondents' (farm principals) replies may have been more influenced by what they regarded as being economically worthwhile and not by what their successors might be prepared to accept. The two upland study areas held the lowest proportions of positive responses, with only 41% of Staffordshire farmers expecting a succession compared with 57% in Dorset and Bedfordshire. Staffordshire and the MGB (Metropolitan Green Belt) held the highest proportion of farmers where it had been definitely ruled out, reflecting low farming incomes in Staffordshire and the more unstable land-occupancy conditions and non-agricultural opportunities displayed in the MGB.

Expectation over succession, while the most important single indicator of family continuity, needs to be qualified by other kinds of information. For this reason farms were allocated to one of five categories on the basis of answers received to a set of questions relating to the origins of the family and the future plans the family had regarding continuity. The measure of family commitment would then incorporate *ex ante* as well as *ex post* social factors. The intention was to construct measures of the general 'social trajectory' of each farm family through its attachment to the continued occupancy of the farm. The results could then be applied to the original subsumption typology enabling an analysis of the tendencies between them and level of subsumption. Making an assessment of this kind is problematic given variations in the existing family life-stage at the time of interview. To some degree, this source of potential error is minimized by taking into consideration the length of family occupance and the degree to which the respondents originated from local farming families.

Five categories of family commitment have been defined.

1. *Strongly established families*: families where succession is planned and where the farm is at least in the second generation of family occupancy.
2. *The establishers*: where succession in the future has not been ruled out (the household may not yet contain children or include young children who have yet to decide whether to pursue an agricultural livelihood) on at least a second generation farm; or a first generation farm where a succession is definitely planned; or a first generation farm where a succession has not been ruled out and the farmer belongs to a local farming family network (i.e. where a farmer has moved from the 'home farm' and set up an autonomous business in the locality).
3. *Potential establishers*: where farms are the first generation family farms, the farmer does not belong to a local farm family network, and a succession has not yet been ruled out.
4. *Uncommitted families*: succession has been ruled out, often because the children are unwilling to take over the farm.
5. *Non-family farms*: where a manager, director or land agent is employed to manage the farm on behalf of a company.

The distribution of farms by category for each of the five areas is shown in Table 4.1. Almost two-thirds of the full sample were either 'strongly established families' or 'establishers', with 28% defined as 'uncommitted families'. The first four categories, from 'strongly established families' to 'uncommitted families', reveal a decreasing average farm size across the sample as a whole. The average farm size of 'uncommitted families' is 84.7 ha, compared with 178.4 ha for the 'strongly established' farm families. These results provide some support for previous research (Gasson *et al.*, 1988), which alludes to stronger levels of family commitment on farms of medium to large size. A more detailed examination of the data reveals differences in this tendency between localities. In both upland study areas the average size of 'establishers' is higher than for 'strongly established families', and 'uncommitted families' in Cumbria occupy farms with a mean size close to the average and larger than for the 'strongly established families'. As these and the succeeding results suggest, however, the development of more complex farm business relations does not negate the role of short- or long-term family commitments. As we shall argue below, they become a prerequisite for success on the agricultural treadmill. The category with the largest average farm size was the 'non-family farm.' Despite being a small minority, such agribusinesses represent an emergent tendency in the development of 'fully subsumed' units.

Table 4.1. Social trajectory categories by locality.

	Staffs	Cumbria	Dorset	MGB	Beds	Total	
Category	%	%	%	%	%	No.	%
Strongly established families	28.6	26.1	36.4	29.1	42.5	138	32.6
Establishers	31.4	46.6	28.3	27.9	27.5	137	32.4
Potential establishers	4.3	5.7	3.0	3.5	5.0	18	4.3
Uncommitted families	34.3	20.5	32.3	36.0	17.5	119	28.1
Non-family farms	1.4	1.1	–	3.5	7.5	11	2.6
Total no. of farms	70	88	99	86	80	423	N/A

MGB: Metropolitan Green Belt (London); Beds: Bedfordshire; Staffs: Staffordshire
Source: Field Survey.

Social trajectories and the subsumption matrix

A key question concerns the relations between the social trajectories for the farms and their locations on the subsumption matrix. The data from Table 4.1 are cross-tabulated against the subsumption matrices 'ideal types' in Table 4.2. Commitment to family continuity increases as farms become more subsumed. In the first three ideal type cells (i.e. 'marginal closed', 'transitional dependent' and 'integrated units'), the proportion of farms managed by 'strongly established families' steadily increases. For the whole population (423 farms), 32.6% fall into this category. 'Strongly established families' are under-represented among the 'marginal closed units', and proportionally more significant among the 'transitional' and 'integrated units'. Examining the 'uncommitted families' indicates that family continuity is under greatest pressure on the less subsumed farms. Almost half the 'marginal closed' enterprises have ruled out succession to the next generation, whereas the proportions decline to 22% for the 'transitional' units and 13.6% for the 'integrated units'.

Table 4.2. Number of farms by social trajectory and subsumption ideal types.

Social trajectory category	Subsumption ideal types increasing subsumption →							
	Marginal closed unit		Transitional dependent unit		Integrated unit		Subsumed unit	
	No.	%	No.	%	No.	%	No.	%
Strongly established families	18	22.2	43	36.4	9	40.9	1	16.7
Establishers	20	24.7	47	39.8	6	27.3	1	16.7
Potential establishers	4	4.9	2	1.7	3	13.6	0	0.0
Uncommitted families	39	48.1	26	22.0	3	13.6	1	16.7
Non-family farms	0	0.0	0	0.0	1	4.5	3	50.0
Total	81	100	118	100	22	100	6	100

Source: Field Survey.

Conventional analyses of petty commodity production would suggest that these results are contradictory, but what they expose is the continuing necessity for family-based reproduction if farms are to develop on an agricultural treadmill set by external capitals and state policy (Marsden, 1991). The evidence suggests a complementarity between the social bases of family continuity and succession, and the increasing tendency for external capitals to reduce the independence and autonomy of farm families. The expectation of survival of particular farm families into the next generation appears to increase as farms become more engaged with technological, marketing and credit links, and, generally, as they experience a higher level of subsumption of production relations through more complex labour and farm business structures; and yet, as evidence from the United States quoted earlier suggests, the risk of failure of such farm businesses during periods of severe farming recession can be greater.

Conclusion: Socially Sustaining Commodity Relations

The incorporation of social trajectories into the analysis suggests that key aspects of family relations and market mechanisms are not necessarily structurally contradictory. They can be mutually reinforcing in directing agricultural development in a period of relative stability and prosperity. These arguments support other evidence that identifies the significance of the 'continuity ethic' with the more 'successful' farmers, suggesting that petty commodity production, based largely on family ownership, occupancy and farm labour, may form a social prerequisite for, rather than a barrier to, the penetration of commodity relations by capital in contemporary agriculture. Stability and commitment in family relations are as much in the interests of capital as they are the farm family. It is, perhaps, no surprise to record that, even at the height of institutional investment in UK agriculture by pension funds and insurance companies during the 1970s, these investors were as concerned with the entrepreneurial potential and business acumen of the occupiers and their families of the farms they purchased, as they were of the physical assets of the enterprise (Munton, 1985). It is, therefore, the ways in which these seemingly contradictory spheres are integrated and distributed within and between farm families that are crucial in sustaining socially the form of production (Marsden, 1991).

In addition, the results of this latest analysis allow us to 'unpack' further the concepts of subsumption and commoditization within a national context. If there is indeed an 'end point' to the process of subsumption, whereby fully commoditized production relations are apparent and the farm family has been completely absorbed, these cases are still in a small minority. They may well remain so for as long as farm families are prepared to exploit their own resources and skills in order to maintain some level of control over the occupancy of their farms. It may be, therefore, that it is on those farms that are aligned to the 'transitional dependent' and 'integrated' ideal types that the more significant sorting processes are likely to occur. At the same time the penetration of external capitals continues to intensify but the farming and food policies that did so much to stimulate them begin to fall away. Moreover, our evidence from upland Britain suggests that the farm businesses of lowland Britain may have some considerable way to go before their levels of self-exploitation became unsustainable. The prevalence of 'strongly established' and 'establisher' families among those farms where dependent relations with external capitals and markets are well developed, reveals the real complexity of the interface between the concepts of use and exchange value. It would seem, contrary to Chevalier (1983) and Bernstein (1986), that the prevalence of exchange relations with a host of external agencies and markets does not necessarily diminish the relevance of long-term 'use' values that encompass family and generational time horizons associated with past occupation and prospective continuity (see also Van der Ploeg, 1990).

Indeed, it could be suggested that the establishment of complex exchange relations with external capitals may provide a commodified means to the more socially oriented ends associated with ensuring family continuity and succession. There are, increasingly, considerable risks to the family in these relations, not least in the adequacy of the market relations to deliver the social benefits farm families increasingly aspire to. Nonetheless, the evidence would suggest that the social coordination of exchange and commoditized relations on the one hand, with the longer time horizons of family-based use values on the other, is an increasingly important activity for farm families if they are to maintain continuity. The price of failing is not abrupt extinction through subsumption, but displacement by those families who have been able to synchronize rather better the demands of capital and their own aspirations to land occupancy.

The analysis points to the need, somewhat ironically, for further exploration of our notions of commoditized production relations and the social dimensions of 'use values' among the more highly subsumed farm enterprises. It is quite evident that any notion of 'complete' market integration, based on generalized notions of petty commodity production, adds little to our understanding of the uneven restructuring of agricultural production relations in capitalist economies (see also Long and Van der Ploeg, 1988). The abandonment of such unilinear assumptions of agrarian development are increasingly urgent when confronted by the empirical realities of the uneven nature of British agriculture, especially since the mid-1980s when greater instabilities in agricultural markets and policies have promoted a wider range of farm family responses.

The development of the typology attempted here and, in particular, the links we have sought to make between empirical diversity and theoretical frameworks, have done little more than to establish some consistent and broad parameters upon which to open new investigations of agrarian production relations. Based upon a realist methodology, we have attempted to establish sets of ideal typical, contingent social and economic conditions under which farm families engage in production through their variable interactions with external agencies and markets. These frameworks have allowed the assessment of uneven development of these production relations across five different agricultural areas and between four ideal farm types. The evidence of empirical variation within the ideal types represents some of the problems of relating the abstract to the concrete, but it does not deny the value of the procedure nor denude it of theoretical insight. Instead, the results pose further questions about the assumptions through which marxist political economy has addressed the topic of agrarian development in general and subsumption of the family-based farming in particular.

Acknowledgement

This chapter incorporates the findings from an analysis conducted by James Kneale. The study was supported by the Economic and Social Research Council and the Leverhulme Trust and we wish to acknowledge their generous financial assistance.

References

Bennett, J. (1982) *Of Time and the Enterprise: North American Family Farm Management in a Context of Resource Marginality*. University of Minnesota Press, Minneapolis.

Bernstein, H. (1986) *Is there a concept of petty commodity production general to capitalism*? Paper presented to the 13th European Congress of Rural Sociology, Braga, Portugal.

Bowers, J. (1985) British agricultural policy since the Second World War. *Agricultural History Review* 36, 66–76.

Chevalier, J. (1983) There is nothing simple about simple commodity production. *Journal of Peasant Studies* 10, 153–86.

Crow, G. (1987) Agricultural rationalisation: the fate of family farmers in post-war Britain. Unpublished PhD thesis, University of Essex.

Friedmann, H. (1986) Patriarchy and property: a reply to Goodman and Redclift. *Sociologia Ruralis* 26, 186–93.

Fuller, A.M. (1990) From part-time farming to pluriactivity: a decade of change in rural Europe. *Journal of Rural Studies* 4, 361–73.

Gasson, R., Crow, G., Errington, A., Hutson, J., Marsden, T.K. and Winter, M. (1988) The farm as a family business: a review. *Journal of Agricultural Economics* 39, 1–42.

Harrison, A. (1975) *Farmers and Farm Businesses in England*. Miscellaneous Study No. 62, Department of Agricultural Economics, University of Reading.

Hawkins, E. (1991) Changing technologies: negotiating autonomy on Cheshire farms. Unpublished PhD thesis, South Bank Polytechnic, London.

Hill, B. (1990) *Farm Incomes, Wealth and Agricultural Policy*. Avebury, Aldershot.

Hutson, J.K. (1987) Fathers and sons: family farms, family businesses and the farming industry. *Sociology* 21, 215–29.

Long, N. and Van der Ploeg, J.D. (1988) New challenges in the sociology of rural development; a rejoinder to Peter Vandergeest. *Sociologia Ruralis* 28, 30–41.

Marsden, T.K. (1991) Theoretical issues in the continuity of petty commodity production. In: Whatmore, S.J., Lowe, P. and Marsden, T.K. (eds), *Rural Enterprise: Shifting Perspectives on Small Scale Production*. Fulton, London, pp. 12–33.

Marsden, T.K. and Murdoch, J. (1990) Agriculture in retreat: Implications for the changing control and development of rural land. *Working Paper* 9, ESRC Countryside Change Initiative, University of Newcastle, Newcastle.

Marsden, T.K., Munton R.J.C. and Whatmore, S.J. (1987) Uneven development and the restructuring process in British agriculture: a preliminary exploration. *Journal of Rural Studies* 3, 297–308.

Marsden, T.K., Munton R.J.C. and Whatmore, S.J. (1989) Strategies for coping in capitalist agriculture: an examination of the responses of farm families in British agriculture. *Geoforum* 20, 1–14.

Mooney, P. (1988) *My Own Boss? Class, Rationality and the Family Farm*. Westview Press, Colorado.

Moran, W. (1988) The farm equity cycle and enterprise choice. *Geographical Analysis* 20, 84–91.

Munton, R.J.C. (1985) Investment in British agriculture by the financial institutions. *Sociologia Ruralis* 25, 155–73.

Munton, R.J.C., Whatmore, S.J. and Marsden, T.K. (1988) Reconsidering urban fringe agriculture: a longitudinal analysis of capital restructuring on farms in the Metropolitan Green Belt. *Transactions of the Institute of British Geographers* 13, 324–36.

Munton, R.J.C., Marsden, T.K. and Whatmore, S.J. (1990) Technological change in a period of agricultural adjustment. In: Lowe, P., Marsden, T.K. and Whatmore S.J. (eds), *Technological Change and the Rural Environment*. Fulton, London, pp. 53–80.

Nalson, J. (1968) *The Mobility of Farm Families*. University of Manchester Press, Manchester.

Northfield Committee (1979) *Report into the Acquisition and Occupancy of Agricultural Land*. Cmnd. 7599, HMSO, London.

Reid, I.G. (1981) Farm finance and farm indebtedness in the EEC. *Journal of Agricultural Economics* 32, 265–74.

Rheinhardt, N. and Barlett, P. (1989) The persistence of family farms in United States agriculture. *Sociologia Ruralis* 29, 203–26.

Salamon, S. (1985) Ethnic communities and the structure of agriculture. *Rural Sociology* 50, 323–40.

Shucksmith, M., Bryden, J., Rosenthall, P., Short, C. and Winter, M. (1989) Pluriactivity, farm structures and rural change. *Journal of Agricultural Economics* 40, 345–60.

Van der Ploeg, J.D. (1990) *Labour, Markets and Agricultural Production*. Westview Press, Boulder, Colorado.

Whatmore, S.J. (1991) *Farming Women: Gender, Work and Family Enterprise*. MacMillan, London.

Whatmore, S.J., Marsden, T.K., Munton, R.J.C. and Little, J. (1987a) Towards ypology of farm businesses in contemporary British agriculture. *Sociologia Ruralis* 27, 21–37.

Whatmore, S.J., Marsden, T.K., Munton, R.J.C. and Little, J. (1987b) Interpreting a relational typology of farm businesses in southern England. *Sociologia Ruralis* 27, 103–22.

Whatmore, S.J., Munton, R.J.C. and Marsden, T.K. (1990) The rural restructuring process: emerging divisions of agricultural property rights. *Regional Studies* 24, 235–45.

5

Farm Business Restructuring in the Urban Fringe: The Toronto and Montréal Regions

Pierre Deslauriers, Christopher Bryant and Claude Marois

Agricultural change in the urban fringe has long been a significant research theme among geographers, and much attention has been devoted to agricultural land loss and the degeneration of farm structures. More recently, attention has been directed toward the other forms of farm change there, including farm adaptation to stress (see Bryant and Johnston, 1991, for a review of this literature). However, certain themes remain relatively undeveloped such as farm-level restructuring including farm diversification. In this chapter, a conceptual framework is presented to deal with farm-level restructuring in the urban fringe. Selected aspects are then illustrated for the urban fringes of Toronto and Montréal, the two largest urban concentrations in Canada.

Farm Business Restructuring in the Urban Fringe: A Framework

Four building blocks related to the dynamics of farm change are introduced: (i) the socioeconomic organization of farm production; (ii) the systems of exchange that tie the farm into broader functioning systems; (iii) the formative forces of change in these systems of exchange; and (iv) the types of farm-level responses. Each building block presents certain specificities in the urban fringe, which produce a more diversified restructuring of agriculture.

First, socioeconomic modes of agricultural production vary over time and space in relation to the evolution of the overall socioeconomic system. Malassis (1958), for instance, argued that agriculture in the Western World had gone through a series of evolutionary changes reflecting broad changes in the whole socioeconomic system. Each of his three 'pure' modes of farm socioeconomic organization – subsistence, family farm (artisanal) and capitalistic – could be characterized by distinctive social, economic, technical, financial and behavioural characteristics.

With the rise of industrial society, the capitalistic and industrial mode of production became increasingly important as agriculture became integrated into the agro-industrial complex (Troughton, 1982a, b; Lowe *et al.*, 1990). However, because of time-lags and spatial diffusion processes, farms at a particular time and place present a range of socioeconomic structures. Adjacent to the urban-industrial complex, some conditions favour the transformation towards a capitalistic and industrial mode of production. These include the pull of agricultural labour into non-farm employment and subsequent farm consolidation, both enabling and encouraging mechanization and motorization of agriculture (Pautard, 1965).

With the development of postindustrial society, other forms of socioeconomic organization also find conditions ripe for their development in the urban fringe of major urban concentrations. Examples with strong links to postindustrial society include some types of part-time farming, hobby farming, organic and other forms of 'sustainable' agricultural production, each potentially with its own particular characteristics, relationships and goals.

How do these broad modes of production change and restructure? The second building block of the conceptual framework, systems of exchange, provides a clue (Bryant, 1988; Bryant and Coppack, 1991; Bryant and Johnston, 1991). All economic activities, including agriculture, function within different systems of exchange (for example, produce markets, capital markets) (Fig. 5.1). An activity can be tied into different systems of exchange at different scales; for example, a particular farm may function simultaneously within an international produce system of exchange, a national system for capital access and a local system for farm labour. Each system of exchange comprises a set of nodes; for instance, for the produce marketing system, these include the production unit and its competitors, retail and wholesaling operations and processing industries; interactions between them (for example, flows of produce) use particular transportation and communication media and networks.

Changes in any of these components can affect the production units because they involve changes in the farm's decision-making environment, altering the values placed upon different resources and enterprises. Since different 'production' activities are tied into different mixes of systems of exchange, they may be exposed to different sets of forces and therefore may be transformed in different ways. Changes come either through macrochanges to which individuals react and adjust, or they can also be initiated by individuals through adaptive behaviour that is more clearly entrepreneurial and innovative (Bryant, 1989).

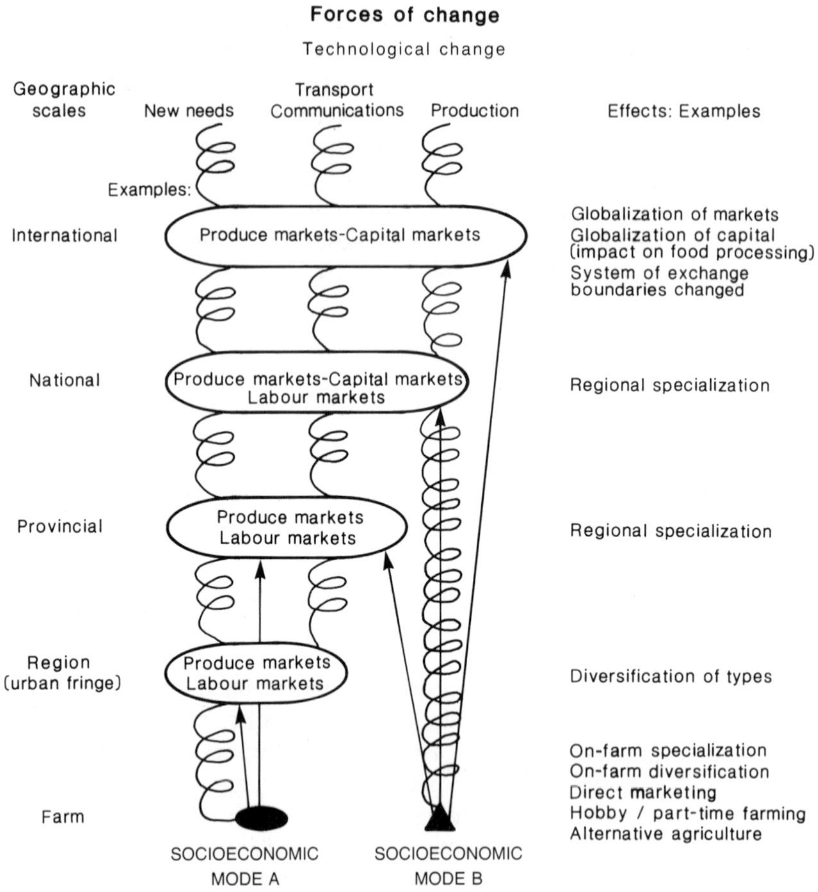

Fig. 5.1. Systems of exchange and agricultural production.

So, farm differentiation in an area may occur because there are different socioeconomic modes of production and because different farms may be tied into quite different systems of exchange. For instance, side by side in an urban fringe environment are farms producing for a national and international system of exchange, as well as farms oriented towards local and regional markets using direct marketing.

The third building block deals with the forces of change. The geographic boundaries of the systems of exchange, as well as the direction and magnitude of exchanges, can be modified in various ways, including:

1. changes in both monetary and non-monetary values (for example, the values attached to different kinds of work, changes in consumer tastes);

2. institutional changes (for example, regulating the operation of systems of exchange through supply management and trade rules); and
3. technological change in production processes and transportation and communications.

Underlying these changes are three sets of forces that characterize postindustrial society: the development of new 'needs', and changing communication and production technology (Bryant, 1988).

'New needs' include the relatively buoyant demand for non-food products, such as horticultural production associated with an urbanizing and more affluent population. Similarly, the increasing interest by the consumer in 'healthy' agricultural products has created a market differentiated by process (e.g. organic produce), showing how non-economic values become integrated into the market place. Other examples include opportunities for farms to diversify into recreational enterprises.

Communication technology has already led to important global transformations of agricultural production. Partly rooted in industrial society, for example the role of transportation in enlarging the feasible supply area for foodstuffs for major markets, it is also partly linked to processes that are firmly embedded in postindustrial society. For instance, the effects of information processing and communication technology include the reorganization of farm management and marketing systems.

Technological change in production processes also has strong roots in industrial society. Much of the 'normal' agricultural change, at least up until the late 1970s, represents the integration of the industrial model of production into agriculture. It includes mechanization, capital substitution for labour, increasing reliance on non-farm inputs and farm business consolidation (see Chapter 2). With postindustrial society, a greater development of alternative technologies supportive of more sustainable agricultural production can be expected (see Part V).

The fourth building block deals with the range of farm level responses. Bryant and Johnston (1991) suggest a three-fold categorization of options followed by farmers in the urban fringe: (i) modification of farm business size (increase or decrease) and enterprise and marketing structure; (ii) moving agricultural labour wholly or partly out of agriculture; and (iii) modification of the organizational structure of the farm, especially its capital structure. Farm business restructuring is more than simply changing technical coefficients; more importantly, it includes transformations of the social and control relationships between ownership and use of the factors of production, of the degree of internalization of farm management and decision-taking, and of the degree of integration of the farm unit into different market structures.

The discussion now focuses on farm business size expansion and diversification. Farm size expansion can involve expansion of the physical acreage of the farm, through purchase or rental of additional land. Farmland rental has played an important role in this process in the urban fringe in several countries (for example, Sublett, 1975; Bryant, 1976). In Chapter 2, Troughton describes this process as part of the general process of the integration of the industrial model into agricultural production. Depending upon circumstances, farmland rental in the urban fringe can represent a very stable relationship or a rather fragile one (Munton *et al.*, 1988).

Farm expansion can also occur through intensification and modification of the marketing channels, and enterprise diversification can sometimes be linked to such marketing changes. Farm diversification both responds to and contributes to the dynamics of restructuring (Ilbery, 1988b), and represents a decrease in the farm family's dependency upon a particular set of income sources.

Diversification can be associated with a net increase in the number of income-generating activities produced on or off the farm using the resources accessible to the farm family, as well as with a net increase in the number of marketing methods used. Diversification in marketing channels represents real diversification because different channels represent different systems of exchange, potentially different market segments and different market behaviour patterns. Included in diversification are both unconventional enterprises and marketing channels, as well as 'traditional' ones. It therefore covers net increases in 'traditional' enterprises, non-traditional enterprises (for example, horticulture) and marketing channels, the incorporation of non-agricultural enterprises on the farm that use redundant or underused farm resources (for example, farm vacations, farm buildings for long-term storage), adding value to farm produce on the farm, and the participation of farm family labour in off-farm work. Diversification, therefore, represents a modification in the internal and external relationships of the farm.

Farm Business Restructuring: Selected Aspects from the Toronto and Montréal Urban Fringes

Farm restructuring in the urban fringe

The research literature on farm-level restructuring in the urban fringe of Canadian cities, or indeed of other cities, is sparse. However, a general picture of agriculture in Canada's urban fringe can be painted. In reviewing selected aspects of agricultural change around Canada's Census Metropolitan Areas, Bryant *et al.* (1984) showed that fringe areas contained high proportions of the most intensive agricultural activities, especially specialized crops, nurseries and horticulture, even though extensively produced crops remained dominant in terms of acreage. These areas also had a larger share of small-size farms and higher levels of farmland rental. Furthermore, fringe farms had the highest level of capital intensity per unit area for land and buildings as well as machinery, and high levels of part-time and hobby farming (Bryant, 1976).

These characteristics, as well as others relating to the personal traits of farmers (age, mode of ownership), farm income and labour, constitute the most frequently noted indicators of farm restructuring. Most of the recent data collected for the Toronto (1985: 189 farm surveys) and Montréal (1988–1989: 149 farm surveys) fringes (Fig. 5.2) show that, overall, the profile of agriculture there conforms generally to trends described 15 years ago (Bryant, 1976). However, the situation is becoming increasingly complex.

Farm expansion is an oft-cited indicator of restructuring in agriculture. As elsewhere, many farmers in Canada have had no choice but to increase production and productivity in order to maintain net farming income levels. This has usually gone hand in hand with expansion of the farm's land base, and the adoption of more capital-intensive farm technology. In the Montréal and Toronto study areas, although small farms (<50 ha) still predominate, there is a tendency for growth in the number of larger farms (>150 ha). There was also an increase in the average size of farms in the Toronto (Johnston, 1989) as well as Montréal areas over the late 1970s and 1980s, mainly due to expansion of farms at the upper end of the size continuum.

In the Montréal study area, farmland rental as a means of expanding the land base was most significant in the most urbanized zones. Inner fringe zones had a higher proportion of farms renting land and of farmland rental than the more remote rural areas. Moreover, this percentage, as well as the number of renters, increased between 1981 and 1988, showing that farmland rental was still progressing (Bryant *et al.*, 1991). In the most urbanized zone of the Montréal study area, 52% of farms expanding between 1981 and 1988 did so through rental, compared with 44% in the more rural zone. Bunce (1985) also observed a link between large-scale, mainly cash-cropping operations, and farmland rental in the Toronto urban fringe, as these farmers fought to attain scale economies.

Around Montréal, intensive crops covered a higher percentage of the total farm area of the fringe areas studied both in 1981 and 1988; this percentage had increased (7.8 to 9.5%) while it had not progressed in the peripheral rural zones. Similarly, in 1985, Bunce reported that as much as one-third of the farms in three municipalities immediately adjacent to Metropolitan Toronto were speciality enterprises typical of the urban fringe, such as pick-your-own vegetable farms, turf farms and horse-riding stables.

Little direct information is available on the investment behaviour of urban fringe farmers. Thus, we must again rely on data collected around

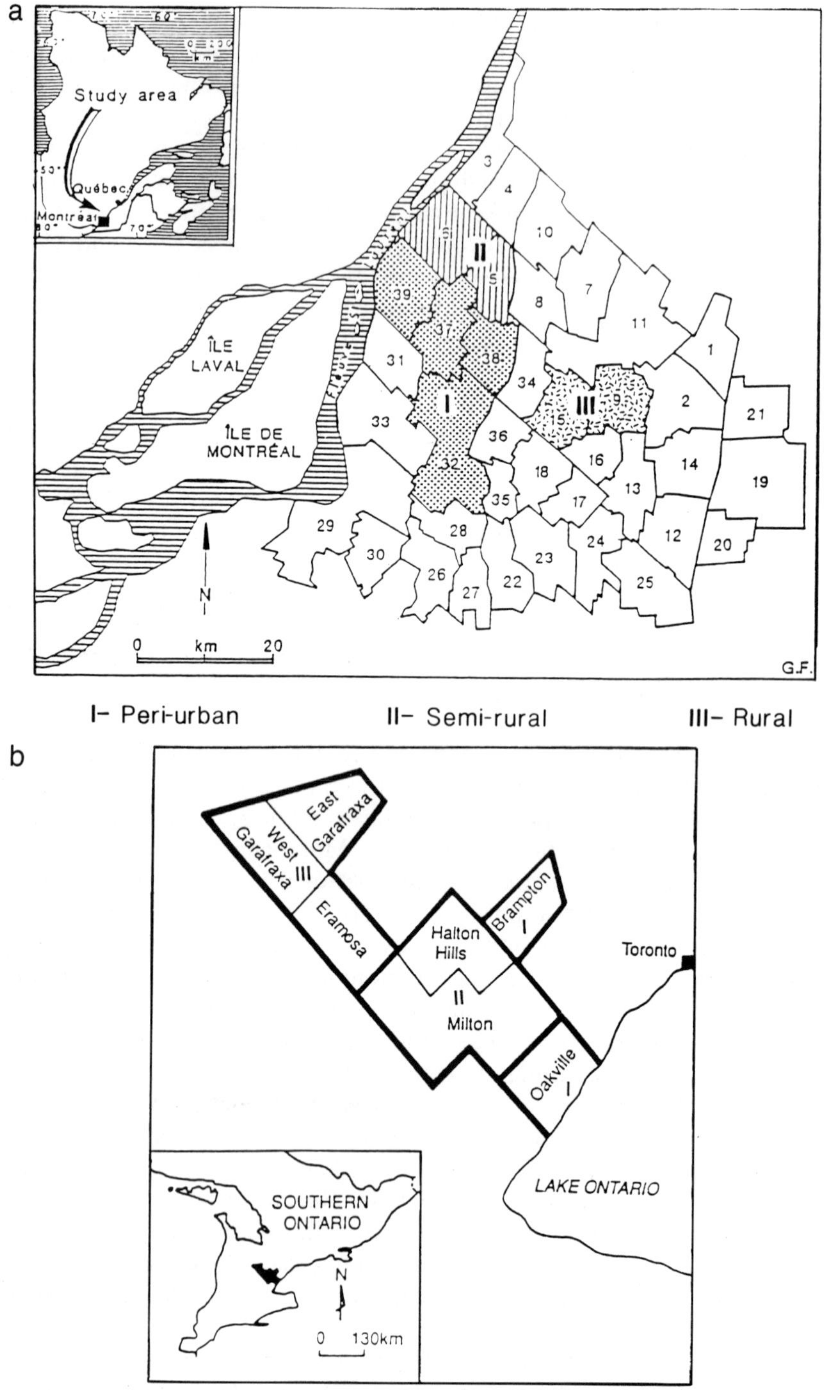

Fig. 5.2. The Toronto and Montréal study areas.

Montréal and Toronto. In the Montréal region, fringe farmers on average had invested more money into various installations and machinery than those in the more peripheral zones; the average amount was also above that for the whole study area. It is noteworthy, however, that the most urban area had the highest percentage of farms, usually small-scale operations, that had made no investments at all.

Capital intensity, measured in terms of dollars invested per hectare in machinery and installations, was higher both in the Montréal fringe and rural areas than the corresponding Toronto areas (for example, a higher proportion – 89% versus 68% – of farms in the more rural zone of the Montréal region showed assets above C$20 000 per ha). There appears to be a tendency for the survival of more 'less capital-intensive' operations in the Montréal fringe, whereas the strikingly high presence of highly capitalized, mostly extensive cash-crop farms is symptomatic of a general overcapitalizing problem that has been noted (and criticized) elsewhere (Caldwell, 1988).

Finally, in terms of farm operator status, there are proportionately more part-time farmers in the Toronto than in the Montréal area. In Montréal, the highest incidence was in the most urbanized zones, whereas in Toronto they were located in a 'transition zone' between the most and least urbanized areas. Surprisingly, the general tendency in both these areas was for a decrease in off-farm employment. While the level of off-farm employment was highest in the zones previously identified as having the most part-time farmers, full-time jobs off the farm seemed on the way to being replaced by part-time off-farm work. Splitting time and income sources more evenly between on- and off-farm activities appeared to be an increasingly popular alternative, reflecting probably as much a search for a life-style change as for any strictly economic benefits. This phenomenon, and the survival of many smaller farms, shows a diversity of socio-economic structures. It suggests a more complex farm structure than the extreme polarization that might be expected from the common process of restructuring in which a polarization appears between farms moving towards the industrialized model of farm production, and those at the other end of the scale which became more and more marginalized. This brief review suggests that although many farms in the Canadian urban fringe show signs of being engaged in the restructuring process involving the industrialization of agriculture, many others have followed different paths.

Farm diversification in the urban fringe

Diversification as one form of restructuring is increasingly recognized as a viable alternative, especially for farmers who seek to survive in urbanized settings without necessarily having to increase their scale of operations significantly. With the multiplication of marketing channels and the adoption of non-traditional enterprises as ways of achieving diversification, farms become tied into different, alternative systems of exchange, achieving greater stability and security of income. Many of the enterprises involved tend not to be in the regulated sectors, thus putting them beyond, at least in part, the institutional controls that protect income for many farmers, but which may also impede the dynamic behaviour of others.

There is some similarity between the Toronto and Montréal regions in terms of diversification. In both, proportionately more farms have engaged in diversification closer to the city than in the more distant 'rural control zones'. When marketing strategies alone are considered, proportionately more farmers were engaged in new strategies around Montréal than in the Toronto study area. More detailed information available for Montréal shows that among those strategies identified as being the main means of diversification, off-farm work is clearly the most significant.

Comparison of the evaluations of the importance for the farm of off-farm income reveals a striking difference between Montréal and Toronto. In the Toronto area, respondents considered this income as less vital to the present and future survival of their farm than in the Montréal area. It was in the more rural zones of those cities' urban fields that this income was considered most needed to support the farm operations, thus confirming the difficult economic circumstances faced by the often over-capitalized, large-scale cash-cropping farms in those zones.

Direct sales and 'horticulture'-type activities are more characteristic of smaller-scale operations in the inner fringe. In the Montréal region their number remained in fact quite small, but nevertheless increased between 1981 and 1988. Interestingly, some farms operating in non-traditional sectors had made some of the largest capital investments, developing into diversified complexes of agricultural, commercial and leisure-oriented activities. For some, this new commercial orientation has led to adding value (food transformation and processing, crafts) to some of their products, whereas others have become engaged in recreational activities.

A series of analyses were undertaken to test relationships between indicators of farm restructuring and diversification behaviour for the Montréal study area. Few statistically significant relationships (using chi squared) were found, but a few are worth noting. When all forms of diversification were considered (including off-farm work, and the addition of all new enterprises), a fairly significant (P=0.076) relationship was obtained for the whole Montréal sample with the variable 'type of farm'; in effect, there was a strong representation of 'other farms' (other than dairy and cash-crop/hay) among those that had diversified. Diversification was also associated strongly (P=0.003) and positively with part-time operators. Also noteworthy was a clearly negative and significant relationship (P=0.003) with the Index of Investment in Machinery.

When diversification through the addition of non-traditional activities only was analysed, negative significant relationships were observed with Growth in Sales of Crop Products (P=0.019), and the Index of Investment in Machinery (P=0.0001), indicating that the diversifying farms were neither those that showed strong growth in sales of crop products, nor those that made significant investments in machinery.

The thin evidence we possess so far thus indicates that, around Canadian cities, most of the operators who have recently chosen to diversify their farm structure are not in the 'mainstream' sectors of agriculture. They are predominantly owners of small and medium-sized farms, who consider themselves part-time, and who do not operate within a highly capitalistic mode of organization. Since they are not subject to the economic imperatives of the land and capital requirements of this latter mode, and tend not to be integrated into regulated sectors, it would seem they are more flexible in their choice of activities. This suggests that they can integrate more readily into smaller scale, less-controlled systems of exchange, through which they can respond to the rapid changes that are occuring in many sectors of society. Under such conditions, individual behaviour acquires increased significance in the evolutionary process of urban fringe agriculture.

Conclusion

Recent findings in urban fringe research show that the progression towards the industrial model is not the sole avenue for agricultural restructuring. At the scale of the whole agricultural system, many commentators suggest the dominant tendency is towards a polarization of farm structure, with an increasing differentiation between large-scale agri-business type operations and small farms, with medium-sized farms finding themselves in an increasingly precarious position. This has also been suggested for some urban fringe environments (see Ilbery and Barrington, 1986). In the urban fringe especially of the major metropolitan areas, however, the situation appears to be much more diversified and complex. Although Smith's (1987) distinction between farms that operate within a 'production mode' and those that choose to function according to a 'value mode' is consistent with the evolving conceptual framework presented above, we would argue for an even greater range of socioeconomic structures of production.

Setting aside off-farm work as a form of farm family diversification, diversification in the urban fringe tends not to be representative of a large proportion of farms. The family farm remains a major feature of urban fringe farming (Ilbery, 1988a), albeit with a strong tendency to incorporate elements of the capitalistic and industrial mode of farming. However, because of the variety and intensity of the formative forces associated with postindustrial society in the urban fringe, other alternative paths have opened up. These are particularly related to market differentiation occurring within the agricultural marketplace, permitting the development of different regional systems of exchange for farm produce. Direct selling has advanced because of the urban consumer's desire for freshness, quality and price advantages – perceived if not real – and potentially can advance even more with the growing interest in 'healthy' products (Bowler, 1981a; Laureau, 1983; Rickard, 1991). Ultimately, this link between the marketing of process as much as the product could extend beyond current 'organic' or 'biological' products to include products produced under other 'sustainable' forms of agricultural production.

It is suggested that the future stability of agriculture in the urban fringe is very much linked to diversification and the development of different forms of farm businesses. Market development would seem to be a key element in this (Ilbery, 1988b), and there is room for these markets to be developed more proactively. Similarly, there are possibilities for income stabilization through the addition of non-farm enterprises on to the farm compatible with an ongoing farm production system. But to capitalize on the opportunities for this, there would have to be changes in the enabling environment created by municipal planning regulations in order not to squash innovation and creativity. In the words of a French author writing of change in the French countryside (Houée, 1990), the diversity of urban fringe agriculture is like a 'thousand work sites' of social and economic experimentation in which the future is being defined. The challenge is how to encourage and harness this creativity for the benefit of all.

Acknowledgement

The authors would like to acknowledge the financial support of the Social Sciences and Humanities Research Council of Canada towards the research reported in this chapter.

References

Bell, D. (1973) *The Coming of Post-Industrial Society: A Venture in Social Planning*. Basic Books, New York.

Bowler, I.R. (1981a) Self service down on the farm. *Geography* 66, 147–50.

Bowler, I.R. (1981b) Some characteristics of an innovative form of agricultural marketing. *Area* 13, 307–14.

Bryant, C.R. (1976) *Farm-Generated Determinants of Land Use Changes in the*

Rural-Urban Fringe in Canada. Lands Directorate, Environment Canada, Ottawa.

Bryant, C.R. (1988) Economic activities in the urban field. In: Russwurm, L.H., Coppack, P.M. and Bryant, C.R. (eds), *The Urban Field: Essays on Canadian Urban Process and Form III*. Publication 30, Department of Geography, University of Waterloo, Waterloo, Ontario, pp. 57–79.

Bryant, C.R. (1989) Entrepreneurs in the rural environment. *Journal of Rural Studies* 5, 337–48.

Bryant, C.R. and Coppack, P.M. (1991) The city's countryside. In: Bunting, T. and Filion, P. (eds), *Canadian Cities in Transition*. Oxford University Press, Toronto, pp. 209–38.

Bryant, C.R. and Johnston, T.R.R. (1991) *Agriculture in the City's Countryside*. Belhaven Press, London.

Bryant, C.R., Marois, C. and Deslauriers, P. (1991) L'agriculture dans la région métropolitaine de Montréal. In: Vachon, B. (ed.), *Le Québec Rural dans tous ses Etats*. Editions Boréal, Montréal, pp. 215–36.

Bryant, C.R., Russwurm, L.H. and Wong, S.Y. (1984) Agriculture in the Candian urban field: an appreciation. In: Bunce, M.F. and Troughton, M.J. (eds), *The Pressures of Change in Rural Canada*. Geographical Monograph 14, Atkinson College, York University, Toronto, pp. 12–33.

Bunce, M.F. (1985) Agricultural land as a real estate commodity: implications for farmland preservation in the North American urban fringe. *Landscape Planning* 12, 177–92.

Caldwell, G. (1988) Surcapitalisation et idéologie de l'enterprise en agriculture. *Recherches Sociographiques* 29, 349–72.

Houée, P. (1990) Espaces ruraux: entre la décomposition et le renouvellement. *Aménagement Foncier Agricole et Rural* 64, 18–26.

Ilbery, B.W. (1988a) Agricultural change in the West Midlands urban fringe. *Tijdschrift voor Economische en Sociale Geographie* 79, 108–21.

Ilbery, B.W. (1988b) Farm diversification and the restructuring of agriculture. *Outlook on Agriculture* 17, 35–9.

Ilbery, B.W. and Barrington, M.J. (1986) Peri-urban farming: a case study from Coventry's urban fringe. *Journal of the Royal Agricultural Society of England* 147, 42–53.

Johnston, T.R.R. (1989) Farmers' adaptive behaviour in an urbanising environment: Guelph to Toronto area. Unpublished PhD thesis, Department of Geography, University of Waterloo, Waterloo, Ontario.

Laureau, X. (1983) Agriculture péri-urbaine: des entreprises pour demain. *L'Agriculture d'Entreprise* 171–2, 3–42.

Lowe, P., Marsden, T.K. and Whatmore, S.J. (eds) (1990) *Technological Change and the Rural Environment*. Critical Perspectives on Rural Change Series I, David Fulton Publishers, London.

Malassis, M. (1958) *Économie des Exploitations Agricoles: Essai sur les Structures et les Résultats des Exploitations Agricoles de Grande et de Petite Superficie*, École Pratique des Hautes Études, Paris.

Munton, R.J.C., Whatmore, S.J. and Marsden, T.K. (1988) Reconsidering urban fringe agriculture: a longitudinal analysis of capital restructuring on farms in the Metropolitan Green Belt. *Transactions of the Institute of British Geographers* 13, 324–36.

Pautard, J. (1965) *Les Disparitiés Régionales dans la Croissance de l'Agriculture Française*, Série Espace Économique, Gauthier-Villars, Paris.

Rickard, T.J. (1991) Direct marketing as agricultural adaptation in megalopolitan Connecticut. In: van Oort, G.M.R.A., van den Berg, L.M., Groenendijk, J.G. and Kempers, A.H.H.M. (eds), *Limits to Rural Land Use*. Centre for Agricultural Publishing and Documentation (Pudoc), Wageningen, The Netherlands, pp. 77–88.

Smith, S.N. (1987) Farming near cities in a bimodal agriculture. In: Lockeretz, W. (ed.), *Sustaining Agriculture Near Cities*. Soil and Water Conservation Society, Ankeny, Iowa, pp. 77–90.

Sublett, M.D. (1975) *Farmers on the Road*. Research Paper 168, Department of Geography, University of Chicago, Chicago.

Troughton, M.J. (1982a) Process and response in the industrialization of agriculture. In: Enyedi, G. and Volgyes, I. (eds), *The Effects of Modern Agriculture on Rural Development*. Pergamon Press, New York, pp. 213–27.

Troughton, M.J. (1982b) *Canadian Agriculture*. Geography of World Agriculture 10, Research Institute of Geography, Hungarian Academy of Sciences, Akademiai Kiado, Budapest.

6

Farm Diversification in the United States

Darrell Napton

The demand for US farm commodities is satiated, and the federal government will no longer guarantee crop support payments or income enhancements. The old rules used by farmers to decide what to grow and how to market it no longer apply; but the new rules are not yet clear. Farmers, agricultural professionals, and governments at every scale are responding with a host of new ideas to promote the economic health of American farm families and their communities, and the agribusiness sector of the economy in general. Agricultural diversification is one path to economic well-being that is under development.

Farm Diversification in Perspective

Diversification has historically been a farm response to risk and uncertainty (Hart, 1975; Anosike and Coughenour, 1990), but in recent decades US farmers have largely abandoned it in favour of technological and institutional risk-reduction techniques. The new techniques were appealing because they supported and encouraged the vertical integration and specialization that seemed to provide hope for farm survival and success. These risk-reduction strategies were expensive and often substituted energy for risk. Irrigation systems, grain driers, refrigerated transportation equipment and biocides reduced the risks of loss to unpredictable natural forces. Crop insurance and emergency government assistance during times of environmental catastrophies also encouraged farmers to specialize without undue concern for risk. These strategies do not seem to be as appropriate or successful under current farming conditions.

Several powerful forces have caused farming to topple into a pit of uncertainty that has increased economic risk and instability (Ayer and Campbell, 1987), even as environmentally caused risk became manageable. Nevertheless three of these forces have been instrumental in providing hope that farmers could diversify to achieve more certainty and higher incomes.

Agricultural success and the world market

Chronic farm surpluses caused by improvements in agricultural technology, coupled with increased integration of the United States into the international economy, have provided a changed agricultural environment (Nothdurft, 1986; Eisinger, 1988). American farmers grow more corn, wheat, soya beans, and other crops that the international market can absorb (Drucker, 1986). Other developed countries are producing their own agricultural surpluses, and developing nations are either producing a surplus of food and fibre or are following a plan leading to self-sufficiency. A low international demand for American farm products, coming at the same time as reduced government subsidies for many major commodities (Nothdurft, 1986) have led to reduced farm income and fewer farmers. Projections indicate that new technologies will further increase crop yields causing production to continue outstripping market demand for many years (Phillips and Sundquist, 1987). Imaginative farmers have started to search for new markets that are not saturated, even if it means growing different crops.

Changing patterns of food consumption

American eating habits have been changing. Concerns about health and diet have prompted much of the change, and the new patterns have fragmented the mass consumption food market (Meeks, 1986). This has provided new niche and speciality markets for some farmers who chose to grow crops to meet the new demands (Meeks, 1987).

The National Research Council and the American Cancer Society are two of several respected organizations that have urged Americans to eat more fruits and vegetables (Shortridge and Shortridge, 1989). Some Americans responded by significantly changing their eating habits. Restaurants now commonly have salad bars, and organically grown products are increasingly popular. The per capita consumption of fruits and vegetables increased by 12% from 1972 to 1987 (Greene, 1988). During the 1980s, most Americans changed the basis of their supermarket selection from the quality of the meat department to the quality and variety of items offered in the produce department (Vietmeyer, 1988). The average number of produce items carried by supermarkets has increased from 65 to 140 (Vietmeyer, 1988). Today some produce departments offer over 250 items (Greene, 1988) with more than 400 different fruits and vegetables available during a year (Vietmeyer, 1988). The consumption of red meat has also decreased in response to health concerns, whereas poultry and seafood consumption has increased (Harvey, 1988; Babb, 1990). As the proportion of older Americans increases, diet and health concerns may well continue to affect food consumption.

Migrants from Mexico, Central America, the Caribbean and Southeast Asia are also affecting food consumption. These people desire familiar foods. First offered in ethnic restaurants and markets, these foods may become as popular as avocados, artichokes or chillies, which were similarly introduced (Vietmeyer, 1988). Today mainstream grocery stores often have ethnic food sections.

The changing work force is also contributing to new eating patterns. As the baby boom generation ages, it also moves up the income ladder. Higher income individuals traditionally are more likely to eat gourmet and exotic foods, and to eat out frequently. People eat differently in restaurants from home, and Americans are demanding different restaurant fare today than they did a generation ago. Restaurants are offering new kinds of foods to meet this demand. Higher incomes and increased international business ventures both lead to more travel abroad where Americans often learn about new foods (Greene, 1988), creating a greater demand for them at home.

The entrepreneurial state emerges

State-level economic policy underwent a historic shift after the mid-1970s (Eisinger, 1988). Federal aid to states began to decline when federal economic policy shifted to the supply-side ideas of deregulation, privatization and the free market. Aggressive international competition led to local economic decline, reduced profits, and the movement of many factories to more profitable overseas production sites. This competition, and an increasingly large negative trade balance, limited the amount of mobile capital that states might attract from other regions of the nation and world.

These new elements of the economic landscape caused states to rethink their historic role of providing public support for private, industry-guided, economic policies. States started to assume larger and more aggressive economic development roles by turning their attention to programmes that emphasized initiative, intervention and guidance. States began nurturing local businesses to generate economic growth and more taxes, rather than competing with other states for limited mobile capital. These new entrepreneurial development policies have given states the tools to pursue economic development by focusing on the demand side of economic growth.

Federal farm programme reductions and state agricultural budget cuts were concurrent with declining farm exports (Nothdurft, 1986) and the resulting farm crisis of the early 1980s. Many rural areas were crippled and the economies of farm states were threatened. The message of the farm crisis was that traditional commodity-oriented farming was endangering the economic viability of communities throughout the nation.

Agricultural Diversification

Two general types of agricultural diversification are emerging in the US to help farmers adapt to the changed agricultural environment: namely entrepreneurial and industrial diversification. Entrepreneurial diversification weds the tools of service delivery and marketing to farming; entrepreneurial farmers find or develop new markets and fill them with specialized, often high-value, products. Industrial diversification is supported by agribusiness and the land grant system; agricultural specialists are developing crops for industrial feedstocks and new products from crops that have recently been overproduced. Each type of diversification is designed to reduce risk and increase farm income, but there are significant differences in their proponents, methods, appeal and geographical impact.

Entrepreneurial diversification

Provoked by the farm crisis and attracted by the new entrepreneurial spirit, some state governors began to view agricultural development as an opportunity to expand and diversify their state's economy. The goal of state-supported entrepreneurial diversification is to lessen the state's dependence on a few traditional crops and to boost the entire agricultural economic sector by import substitution, value-added processing, local job growth and tax roll enhancement. State departments of agriculture that had been regulatory agencies with loyalties to traditional crops and production methods were suddenly encouraged to be creative and find or develop new markets for local agricultural products (Nothdurft, 1986). New ideas were essential. The framework for this was state entrepreneurialism. The thrust was more effectively to match state resources with the market (Green, 1988). State support for entrepreneurial diversification is often based on the assumption that the current farm infrastructure is suitable for an alternate crop mix and that the experience and skills of commodity farmers can be transferred to different crops (Nothdurft and Popovich, 1991). The more aggressive states view their role as a provider of leadership and assistance, and as a catalyst for public and private cooperation that will lead to diversification. They connect entrepreneurs with new ideas and technologies and the resources needed to make the ideas work (Texas Department of Agriculture, 1986). Specifically, entrepreneurial states conduct marketing studies, introduce inquiring farmers to experts who can assist them, operate demonstration farms, conduct research, publish informational literature, and invest in selected diversification schemes that promise long-term payoffs.

Some of the strongest diversification support has come from *Successful Farming* magazine, a farm periodical historically devoted to conventional agriculture (Meeks, 1987). *Successful Farming* legitimized entrepreneurial diversification by sponsoring two conferences that each profiled 100 successful, diversified, entrepreneurial farmers (*Successful Farming*, 1986, 1987).

Entrepreneurial farmers find or develop new markets that result from changed consumption patterns. The general product types to fill these niches include fruits and vegetables, horticultural crops, livestock, field crops, services and recreation, aquaculture and biotech products for scientific uses. Entrepreneurial diversification generally refers to the production of alternative crops, or those crops that are not traditionally grown or exploited in an area. Alternative crops may range from blueberries and emus to catfish and pinto beans, and from ginseng to herbs and mushrooms. Historically some alternative crops have become established as major traditional crops, whereas others never achieved the potential that their promoters had hoped. At one time, soyabeans, peanuts, potatoes, and tomatoes were alternative crops (Meeks, 1986; Lockeretz, 1988), so were the less successful tung oil and Jerusalem artichokes (Paarlberg, 1988).

Individual entrepreneurs may try to diversify so that they can continue to farm. During the farm crisis of the 1980s, some bankrupt farmers switched to alternative crops that required fewer hectares to make a living than traditional commodity crops (Meeks, 1987). The fragmentation of the mass food consumption market provided them with alternatives to produce niche goods (Nothdurft and Popovich, 1991). Many solvent farmers also sought opportunities to diversify their farm operations. Some wanted a supplementary or alternate income that preferably ran in counter cycle to current crops (Nothdurft and Popovich, 1991). Some of these farmers also sought to gain more control over their destiny than participating in federal programmes and traditional commodity farming allowed them (Nothdurft and Popovich, 1991).

Entrepreneurial diversification is market driven rather than production driven (Nothdurft, 1986). Individual farmers prepare market analyses to find profitable market niches and experiment with growing non-traditional crops to fill them. Their efforts are likely to focus on small, clearly defined markets. The thrust is to match more effectively the farmer's resources with the market (Green, 1988).

Entrepreneurial farmers share some characteristics that distinguish them from many other farmers. Entrepreneurial diversification appeals more to optimists and risk takers even though it is a strategy to lessen risk (Nothdurft and Popovich, 1991). Most farmers are conservative and see switching to untried crops as a foolhardy venture, especially if it requires a local infrastructure that is minimal or non-existent (Meeks, 1986). Switching to new crops may produce initial losses even for the best farmers, and there is no guarantee that the venture will ever succeed (Hart, 1975). Deeply indebted farmers would have to be wary of ideas that practically guarantee large initial capital outlays followed by years of losses and uncertainty. Diversification that does not have the barriers of risk, new knowledge and capitalization, carries the risk of quick market saturation and rapidly declining profits. Entrepreneurial diversification may also require skills that are not transferable from conventional farming (Babb and Long, 1988). Entrepreneurial diversifiers, for instance, often concentrate on selling services to supplement, or in place of, physical products. A catch-your-own fish farm, for example, provides the experience of catching the fish in addition to eating it. Many entrepreneurial diversifiers are new or part-time farmers (Babb, 1990) seeking to find a successful niche that will accommodate their land, work schedule or interests. They realize that it would be impossible to compete with large, monoculture farming specialists. In Central Texas, for example, Haskel Griffin, a former house contractor, has shifted his focus from growing subdivisions to growing organic tomatoes that he markets through a farm-to-consumer club (*Austin American Statesman*, 1991). Griffin is the first Central Texas farmer to use this marketing method, which provides in-season vegetables to members to pay a monthly fee for fresh produce. Griffin has transferred his house-building customer skills to the new enterprise. He explained that, 'You find out what the customer wants and how to give it to them efficiently, in a cost-effective way' (*Austin American Statesman*, 1991).

Entrepreneurs may turn to their family, community or state departments of agriculture and commerce for assistance, but the support structure for entrepreneurial farmers is limited. The land grant system that is available to help traditional commodity farmers is reluctant to assist entrepreneurial farmers, because its scientists, administrators and agents are largely ignorant of, and unenthusiastic about, diversification projects that are small, risky and may be opposed by its traditional commodity constituents (Nothdurft and Popovich, 1991). On the other hand, some non-traditional organizations, universities and entrepreneurial states have started to assist farmers who want to diversify. In Cedar Rapids, Iowa, Kirkwood Community College has established a Rural Development Center that provides step-by-step assistance for farmers who want to find new rural business options (Kirkwood Rural Development Center, undated). The college has successfully helped entrepreneurial farmers diversify into fresh vegetables, dried floral arrangements, rabbits, fish, fallow deer and a bed-and-breakfast service (Kirkwood Rural Development Center, 1990). The Northwest Area Foundation has funded several entrepreneurial efforts throughout the Upper Midwest, including one that specializes in lean, drug-free pork, and another that provides farm-fresh chickens. The foundation has also sponsored the Minnesota Food Association's attempt to establish a network of local food growers and processors. The University of Missouri has spearheaded a cooperative effort with several state and federal agencies to establish the Missouri Alternatives Center to provide crop and business information to individuals who want to produce alternative crops (Missouri Alternatives Center, 1990). Some communities have supported entrepreneurial diversification when local marketing and processing could increase community economic stability by providing jobs (Nothdurft and Popovich, 1991).

Entrepreneurial farmers have three major concerns to contend with after they become established. First, diversification may not lead to risk reduction. Vegetables, for instance, have a greater variability in returns than cash grain (Weimar and Hallam, 1988). Second, many entrepreneurial farmers have specialized in small markets that can easily become saturated. Third, entrepreneurial farmers have committed themselves to producing for the market at a time when local and global markets are changing rapidly. Markets are fragmenting, consumer tastes are changing, and foreign competition is unpredictable.

Industrial commodity diversification

Industrial commodity diversification is better organized, yet not as fully developed as entrepreneurial diversification. It is the result of federal, land grant and agribusiness efforts to develop industrial feedstock crops to supplement traditional grain and fibre production. Farmers will be invited to participate after most questions have been answered and markets prepared. This strategy may help overcome the initial caution of commodity farmers. Industrial commodity diversification is not targeted toward market niches that may have room for only a few newcomers. It is more likely to devote attention to industrial feedstocks to produce the raw material for fuels, synthetics, lubricants and fibres (Paarlberg, 1988). Industrial feedstock crops include new crops such as canola, rapeseed, jojoba, guayule and kenaf, as well as traditional crops such as corn and soyabeans from which new products are being developed.

Commodity surplus farmers, the most likely group to participate in industrial diversification, are more often conservative, conventional, continuing farmers with an emphasis on growing corn, cotton, soyabeans and wheat. Their support structure includes the United States Department of Agriculture, the state–federal Land Grant system, and conventional agribusiness suppliers, marketers and processors. This infrastructure emphasizes traditional crops and markets (Meeks, 1986) or carefully selected industrial feedstock alternatives.

Industrial commodity diversification is national in scope (Table 6.1). The intent is to provide new crops for large agricultural regions, because these areas have chronic economic problems. The goal of industrial commodity diversification is to improve farm income and promote rural economic development by reducing commodity surpluses and exporting the new crops to improve the national balance of trade (Thompson, 1988). Some of the new crops are to help the US develop strategic reserves of some items needed for national defence and to provide industrial feedstocks and alternative animal feed (US Senate, 1987; Thompson, 1988; Dicks and Buckley, 1989). Industrial diversification should also reduce risk by focusing on crops and varieties that are less vulnerable to, or targeted toward, adverse and changing environments.

Table 6.1. Regions where selected industrial crops could be grown.

Crop	Location
Kenaf	South of 30°
Rapeseed	Upper South centred on Ohio–Mississippi confluence; Pacific Northwest
Canola	Corn Belt, Great Plains, Pacific Northwest, Southeast and Upper South
Guayule	Desert Southwest
Lesquerella	Southwest
Jojoba	Sonora Desert of Arizona and California
Meadowfoam	Alaska, Pacific Northwest, Eastern US
Crambe	Midwest, Great Plains, West Coast, Southwest

Source: compiled from Dicks and Buckley (1989).

Impact of Diversification on US Agricultural Geography

The income from new crops and associated national exports could total US$15–20 billion, or about 10% of 1988 farm receipts (Dicks and Buckley, 1989). In selected regions, such as the Mississippi Delta where aquaculture could have a major impact, local processing may increase employment and slow out-migration, but in most rural areas impacts will be limited.

Entrepreneurial diversifiers are having their greatest impact in urban markets where they provide a greater variety of locally produced, fresh food than had previously been available. Since entrepreneurial diversification is market driven and market oriented, these farmers are generally located nearby, often in or near the urban fringe, where they have advantages over more distant farmers (Babb and Long, 1988). Metropolitan farmers already specialize in growing fruits, vegetables, and nursery crops on small, intensively tilled farms; many of these farmers are already accustomed to dealing with customer relations and market analysis in addition to selling their product. Becoming more service-oriented by adding a hay ride or petting zoo may not take a significant adjustment for them. Metropolitan farmers have additional advantages because their fruit and vegetable focus has allowed them to remain free from dependence on government commodity and income support programmes, and allowed them to grow what they choose rather than obligated them to grow crops that are not in demand.

Entrepreneurial farmers are information specialists. Their success depends on their knowledge of and ability to take advantage of local variations in soils, slope and climate. As their production becomes increasingly market driven, they must also become knowledgeable of changing tastes, market segmentation, advertising and other business tools to acquire, manage and maintain access to local markets. Urban fringe farmers generally have a comparative advantage in acquiring market access, but not all entrepreneurial farmers who produce high-value products are located near their market. Ostrich ranchers, for example, can sell ostrich eggs, feathers, leather, meat and breeding stock at premium prices regardless of their location.

The total area and number of farmers involved in entrepreneurial diversification will probably remain small. Weimar and Hallam (1988) concluded that 75 farms of 40 acres each (16 ha) could meet Iowa's need for 13 selected vegetables that could be grown there. Approximately US$17 million could be retained in the state through this import substitution scheme. A few farmers would benefit, but the overall gain would be too slight to encourage farmers to shift from corn and soyabeans to vegetables. If farmers in Iowa's neighbouring states also diversified and provided competitively priced products, the benefits would be even less.

Six million acres (2.4 million ha) are currently used to grow the nation's fruits and vegetables (Babb and Long, 1988). If the market for fruit and vegetable products grew by 5% a year for 15 years and imports remained constant, the land needed would increase to only 12 million acres (4.9 million ha). This contrasts with 330 million acres (134 million ha) currently devoted to conventional crops.

The widespread state-level interest in agricultural diversification threatens to make the search for new crops competitive between different regions. To prevent a zero-sum game from emerging, rural development must focus upon developing new markets (Castle, undated). The emerging Texas blueberry industry illustrates these ideas. Historically blueberries have not been significant in Texas agriculture. The Texas Department of Agriculture has targeted blueberry production in the acid, sandy soils of East Texas for one of its entrepreneurial diversification schemes for two reasons. First, with over 16 million residents, there is a large and growing local market. Second, blueberries ripen earlier in Texas than in the competing Southern states of Florida, Mississippi and Louisiana. Texas is located between these states, while a large unserved blueberry market exists in the west and northwest of the US; producers hope to use their location relative to the competition to acquire the market (Texas Department of Agriculture, 1986).

The long-term potential for industrial diversification is large if renewable industrial feedstocks become economically competitive. For this to happen, agricultural products will need to begin supplementing petroleum and pulp trees, and new uses must be discovered for corn and soyabeans. In the short-term, industrial commodities will have little impact on the agricultural geography of the nation. The needs of one newsprint plant can be met with only 4500 acres (1823 ha) of kenaf in the Lower Rio Grande Valley, where the nation's only kenaf mill is being completed (Glynn, 1991). Industrial rapeseed and canola will probably be the next two successful industrial commodities. The demand for them can be met by 450 000 acres (182 250 ha) of farmland (Glynn, 1991), or less than 0.2% of the land currently devoted to crops. Industrial commodity diversification will likely have a different spatial arrangement than entrepreneurial diversification. Broad general farming regions that have the appropriate climate for a particular crop will be targeted. The location of the consumer will be less important than for entrepreneurial diversification, because the crops will undergo intermediate processing. Within the broad farming regions where the right conditions are found, diversifiers will cluster near processing plants that will be constructed in areas in which most of the conventional farm infrastructure is in place and can be used with the new crop. In regions distant from cities, traditional commodity farmers may contribute to national diversification efforts by diverting some efforts to new crops to satisfy the nation's needs for renewable energy and industrial feedstocks. Some processing plants may be constructed in marginal farming areas or areas distant from markets to bolster regional economies, such as the processing plant for canola and industrial rapeseed that is being constructed in Vela, North Dakota (Dicks and Buckley, 1989). Plants that produce new products from conventional crops could be located nearly anywhere within the regions where these crops are grown.

Conclusion

The rural and agricultural geography of the United States is undergoing a transition, and any transition to a new farming system is always difficult (Hart, 1975). Both entrepreneurial diversifiers and industrial commodity diversifiers are muddling forward, albeit somewhat differently. Experimentation, trial and error, accumulation of experience and market analysis all must be accomplished before a new crop becomes successful. Entrepreneurial diversifiers often accomplish this themselves with little, if any, help from traditional sources of farm support, while industrial diversifiers wait for the land grant-agribusiness system to prepare the infrastructure and market.

No single new crop is likely to make the agriculture of any state or region substantially more diverse or markedly to increase farm income, and not all farmers are going to begin raising ostriches, crawfish or blueberries. If diversification efforts are successful, however, some pockets of farmers throughout the nation will develop new and perhaps profitable enterprises.

Farm diversification promises an invigorated agriculture for some farmers. Common sense and ecology support increasing farm diversity because monoculture farming has shown itself to harm the ecosystem. Diversification may also lower the risks and uncertainties associated with market fluctuations and overproduction of major food and fibre commodities.

Higher farm incomes and stable, low-risk profits are worthy goals, but to what extent diversification can deliver remains to be seen. It is still too early to measure the impact of diversification on rural land use or to tell whether it can be a successful strategy for more than a few farmers. Nevertheless, this study has identified the forces that will shape the emerging pattern of rural land use. World surpluses of traditional crops, changing patterns of consumption, and the emergence of the entrepreneurial state have set the stage for the development of two types of agricultural diversification in the USA. Market-driven entrepreneurial diversification will most likely affect rural land-use patterns near cities and at the local and state levels, whereas production driven industrial commodity diversification will be more likely to have an effect on regional or national rural land-use patterns. As these two different types of agricultural diversification develop, a new pattern of land use will emerge that will reflect the altered trade, food and industrial needs of the nation.

References

Anosike, N. and Coughenour, C.M. (1990) The socioeconomic basis of farm enterprise diversification decisions. *Rural Sociology* 55, 1–24.

Austin American Statesman (1991) 19 March.

Ayer, H. and Campbell, H. (1987) A risky business: what's to be done? In: *Policy Choices for a Changing Agriculture*, B–1580, Texas Agriculture Extension Service, Austin, pp. 25–30.

Babb, E.M. (1990) Marketing new crops. In: Janick, J. and Simon, J.E. (eds), *Advances in New Crops. Proceedings of the First National Symposium – New Crops : Research, Development, Economics*. Timber Press, Portland, pp. 6–11.

Babb, E.M. and Long, B.F. (1988) Alternative enterprises for strengthening southern agriculture. In: Beaulieu, L.J. (ed.), *The Rural South in Crisis: Challenges for the Future*. Westview Press, Boulder, pp. 344–57.

Castle, E.N. (undated). Policy options for rural development in a restructured rural economy: an international perspective. In: *Agriculture and Beyond: Rural Economic Development*. College of Agriculture and Life Sciences, University of Wisconsin, Madison, pp. 11–28.

Dicks, M.R. and Buckley, K.C. (eds) (1989) *Alternative Opportunities in Agriculture: Expanding Output Through Diversification*. Agricultural Economic Report 633, Commodity Economics Division, Economic Research Service, United States Department of Agriculture, Washington DC.

Drucker, P.F. (1986) The changed world economy. *Foreign Affairs* 64, 768–91.

Eisinger, P.K. (1988) *The Rise of the Entrepreneurial State: State and Local Economic Development Policy in the United States*. The University of Wisconsin Press, Madison.

Glynn, P. (1991) New crops, and old, offer alternative opportunities. *Farmline* 12(1), 12–16; (2), 12–15.

Green, M.J. (1988) *Agricultural Diversification Initiatives: State Government Roles in Rural Revitalization*. Council of State Governments, Lexington.

Greene, C. (1988) A new look for supermarket produce sections. *National Food Review* 11, 1–5.

Hart, J.F. (1975) *The Look of the Land*. Prentice-Hall, Englewood Cliffs, NJ.

Harvey, D. (1988) Aquaculture: meeting fish and seafood demand. *National Food Review* 11, 10–13.

Kirkwood Rural Development Center (1990) *Tomorrow's Harvest*. Kirkwood Community College, Cedar Rapids.

Kirkwood Rural Development Center. Various publications, various dates, Kirkwood Community College, Cedar Rapids.

Lockeretz, W. (1988) Agricultural diversification by crop diversification: the US experience with the soybean. *Food Policy* 13, 154–66.

Meeks G. (1986) *The State of Agriculture: Some Observations*. National Conference of State Legislatures, Denver.

Meeks, G. (1987) *Agriculture, Economics, and Environmental Protection*. National Conference of State Legislatures, Denver.

Missouri Alternatives Center (1990) *Ag Opportunities*. University of Missouri, Columbia.

Nothdurft, W. (1986) *Going to Market: The New Aggressiveness in State Domestic Agricultural Marketing*. Council of State Policy and Planning Agencies, Washington DC.

Nothdurft, W. and Popovich, M. (1991) Bucking the system: lessons from one Foundation's investments in agricultural diversification. *Northwest Report* No. 11, The Northwest Area Foundation, St Paul, MN.

Paarlberg, D. (1988) The economics of new crops. In: Janick, J. and Simon, J.E. (eds), *Advances in new crops. Proceedings of the First National Symposium – New Crops: Research, Development, Economics*. Timber Press, Portland, pp. 23–6.

Phillips, M. and Sundquist, W.B. (1987) A new technical revolution: how will agricultre adjust? In: *Policy Choices for a Changing Agriculture*, B–1580, Texas Agriculture Extension Service, Austin, pp. 13–17.

Shortridge, B.G. and Shortridge, J.R. (1989) Consumption of fresh produce in the Metropolitan Unitcd States. *Geographical Review* 79, 79–98.

Successful Farming (1986) *Proceedings of the Adapt 100 Conference*. Successful Farming. Des Moines, Iowa.

Successful Farming (1987) *Proceedings of the Adapt 2 Conferences*. Successful Farming, Des Moines, Iowa.

Texas Department of Agriculture (1986) *Economic Growth Through Agricultural Development: A Blueprint for Action*. TDA, Austin.

Thompson, A.E. (1988) Alternative crop opportunities and constraints on development efforts. In: Hardman, L.L. and Water, L. (eds), *Strategies for Alternative Crop Development: Case Histories*. Crop Science Society of America and the Centre for Alternative Plant and Animal Products, Anaheim (CA), pp. 1–10.

US Senate (15 May 1987) Alternative Agricultural Products Research Act of 1987. Senate hearing 100–208.

Vietmeyer, N. (1988) Foreword. In: Janick, J. and Simon, J.E. (eds), *Advances in New Crops. Proceedings of the First National Symposium – New Crops: Research, Development, Economics*. Timber Press, Portland, pp. xvii–xxii.

Weimar, M.R. and Hallam, A. (1988) Risk, diversification, and vegetables: an alternative crop for Midwestern agriculture. *North Central Journal of Agricultural Economics* 10, 75–89.

7

STATE-ASSISTED FARM DIVERSIFICATION IN THE UNITED KINGDOM

Brian Ilbery

Agricultural policy in the United Kingdom and the European Community (EC) is being reappraised. Various supply control measures and environmental instruments have been introduced under the Common Agricultural Policy (CAP) since the mid-1980s, as capitalist agriculture moves into a post-productionist era (Robinson, 1991; Symes, 1991). This coincides with growing international pressures, through the GATT negotiations, to make significant cuts in farm subsidies. The British response to these problems has been enacted through the 1986 Agriculture Act and the 1987 ALURE package (Alternative Land Use and the Rural Economy). Under the latter, different measures and financial support were proposed for farm woodland, forestry, farm diversification and an extension of Environmentally Sensitive Areas (ESAs). The number of ESAs has since been increased (Baldock *et al.*, 1990; Brotherton, 1991), and both Farm Diversification and Farm Woodland Schemes were introduced in 1988 (Ilbery and Stiell, 1991).

This major change in emphasis in agricultural policy has coincided with increasing pressures on farm incomes from a cost-price squeeze (Bowler, 1987). With limits now placed on CAP spending, farmers cannot increase their incomes by producing more, often unwanted, food. Consequently, they are having to restructure their businesses and adjust to a new situation by improving efficiency, cutting costs and finding alternative sources of income (Marsden *et al.*, 1986a, 1989). Different *adjustment strategies* are available to farmers (Munton, 1990), but most relate to either the redeployment of on-farm resources or the development of new sources of farm-family income both on and off the farm.

Farm diversification is one possible adjustment strategy, but on its own cannot solve the current problems facing the agricultural industry. Not all farmers will be able to diversify and the potential for diversification will vary spatially, reflecting both market opportunities and the agricultural geography and traditions of different areas. This chapter is concerned with the role of farm diversification in the restructuring process. It is ordered into three parts: the first discusses some conceptual issues; the second briefly explores national agricultural policies in the United Kingdom designed to encourage farmers to diversify their business interests; and the third examines the uptake of two particular UK policy measures, especially the characteristics of 'adopters' and 'non-adopters'.

Conceptual Considerations

Three conceptual issues of relevance to farm diversification and agricultural restructuring can be outlined briefly.

Internal and external farm environments

The use of a broad political economy approach to the understanding of agricultural change has been advocated in recent years (Marsden *et al.*, 1986b, 1987). This philosophy allows a conceptualization of the behaviour of individuals as constrained by the political economy in which such action occurs. Agriculture is, therefore, transformed by wider political and macroeconomic forces, which in advanced industrial societies stem from the capitalist mode of production. Over time, family farms will become increasingly subsumed by outside capitals. Unfortunately, such an approach relegates farmers to the role of non-decision-makers and *constraints* are emphasized at the expense of *choices* (Duncan and Ley, 1982). Thus, the political economy approach needs to be modified to incorporate a behavioural element. Such an element would recognize that individuals are 'active' agents in the shaping of their own destinies (human agency). Indeed, Whatmore *et al.* (1987, pp. 120–21) confirm that in order to understand agricultural restructuring there is a need to examine the individual farm family members as active participants and 'not simply as passive subjects of inevitable structural process'.

Consequently, a modified political economy perspective can be advocated, in which 'environments' both *external* and *internal* to the farm can be identified (Evans and Ilbery, 1989). The former comprises the various institutions and organizations that exert an influence on farms. This influence is manifest as external finance beyond the farm gate, operating independently of particular farms. Three interdependent types of external finance in agriculture have been widely recognized: financial institutions, technological developments and marketing organizations. Both public and private organizations will thus become involved in the agricultural restructuring process.

A similarly crucial internal farm business environment can be identified. This refers to the complex organizational structure of individual farm-family businesses in terms of the capital, land and labour relations operating within farm holdings. It is important, therefore, to appraise the motives and attitudes of farm family members, the social relations between members, and the stage in the overall family life cycle (Harrison, 1975; Gasson *et al.*, 1988; Whatmore, 1991). The role of individual decision-makers is a central constituent in this internal environment category.

Internal and external environments necessarily interact, representing the interface between structure and agency. This *interactive* environment consists of two interrelated parts: first, the external environment will influence the internal through attempted capital penetration (subsumption) of farm business structures; and secondly, particular farm businesses may make responses to these outside influences and implement internal adjustments accordingly. Any set of interactions will vary with the location of the farm business because of geographical and historical specificity.

Diffusion of agricultural innovations

Farm diversification and agricultural restructuring can be viewed as the adoption of an innovation. Two perspectives on the diffusion of agricultural innovations are relevant in this respect. The first is the *demand* (adoption) perspective, which was advocated in the seminal works of Hagerstrand (1967) and Rogers (1962) and is analogous to the internal farm environment discussed above. This assumes that farmers demand new innovations and that patterns of uptake will relate to the relative innovativeness of adopters, which in turn is influenced by a set of farm and farmer characteristics. Initially, therefore, only a few 'innovators' will adopt the innovation; these are the more dynamic farmers, with larger farms, higher incomes, higher educational and training attainments, and higher levels of capital indebtedness. For the majority of farmers, there will be a lack of interest and a genuine resistance to agricultural restructuring, which will be removed only when there is general awareness and acceptance of the basic principles surrounding such concepts as farm diversification.

However, authors like Brown (1981) and Unwin (1988) have emphasized that this is not the first stage in the diffusion process. Such a stage has become known as the *supply* (market and infrastructure) perspective, dominated by the strategies and policies implemented by diffusion agencies in the external environment to induce adoption. This perspective assumes that the ability to adopt an innovation is far from even; it is constrained by the policies of different institutions, which through promotional communications and market segmentation may target particular groups of farmers and types of farm. In this respect, non-adoption may not reflect a lack of innovativeness, but problems of access to resources (for example, information, capital, education). Clearly, a proper understanding of farm diversification and agricultural restructuring can be obtained only by examining the strategies of diffusion agencies and the relative innovativeness of farmers.

Farm diversification

As revealed in Chapters 5 and 6, the concept of farm diversification is not amenable to very precise definition (McInerney *et al.*, 1989). Nevertheless, there is now general agreement that it refers to the development of non-traditional (alternative) enterprises on the farm (Haines and Davies, 1987; Slee, 1987). A contrast with conventional farming is implied and so farmers producing products in a different way (for example, organic farming) or changing to such unusual enterprises as snails, herbs and evening primrose are diversifying. Similarly, farmers adding value through on-farm processing and direct marketing, or developing farm-based accommodation and recreational activities, are also diversifying (Ilbery, 1988).

Farm diversification, therefore, involves a diversion of resources (land, labour and capital) previously committed to conventional farming activities. For the purposes of this chapter, the concept is restricted to farm-centred activities and thus excludes income generated from, and labour involved in, other gainful activities (OGAs) off the farm. The latter relates to the much broader concept of *pluriactivity*, which has been the focus of considerable recent attention (Gasson, 1988; Shucksmith *et al.*, 1989; Fuller, 1990).

It is possible to distinguish two forms of diversification: *agricultural* and *structural* (Ilbery, 1991). The former remains focused on farming but entails a broadened notion of what is meant by 'farm work'; farm woodland, organic farming and unconventional crops and livestock are good examples (Fig. 7.1). The latter is geared outward from the farm and toward the public and thus marketing is as important as production; examples would include farm-based tourism (accommodation and recreation), adding value by either the processing and/or direct marketing of food, and passive diversification, where land and/or buildings are leased for non-agricultural purposes.

The External (Supply) Environment

The year 1988 signalled the introduction in Britain and the EC of national and international policies designed to encourage farmers to retire farmland and diversify their business structures. In the space available, it is possible only to outline the role of one national institution (the Ministry of Agriculture, Fisheries and Food – MAFF) through two of its policy instruments

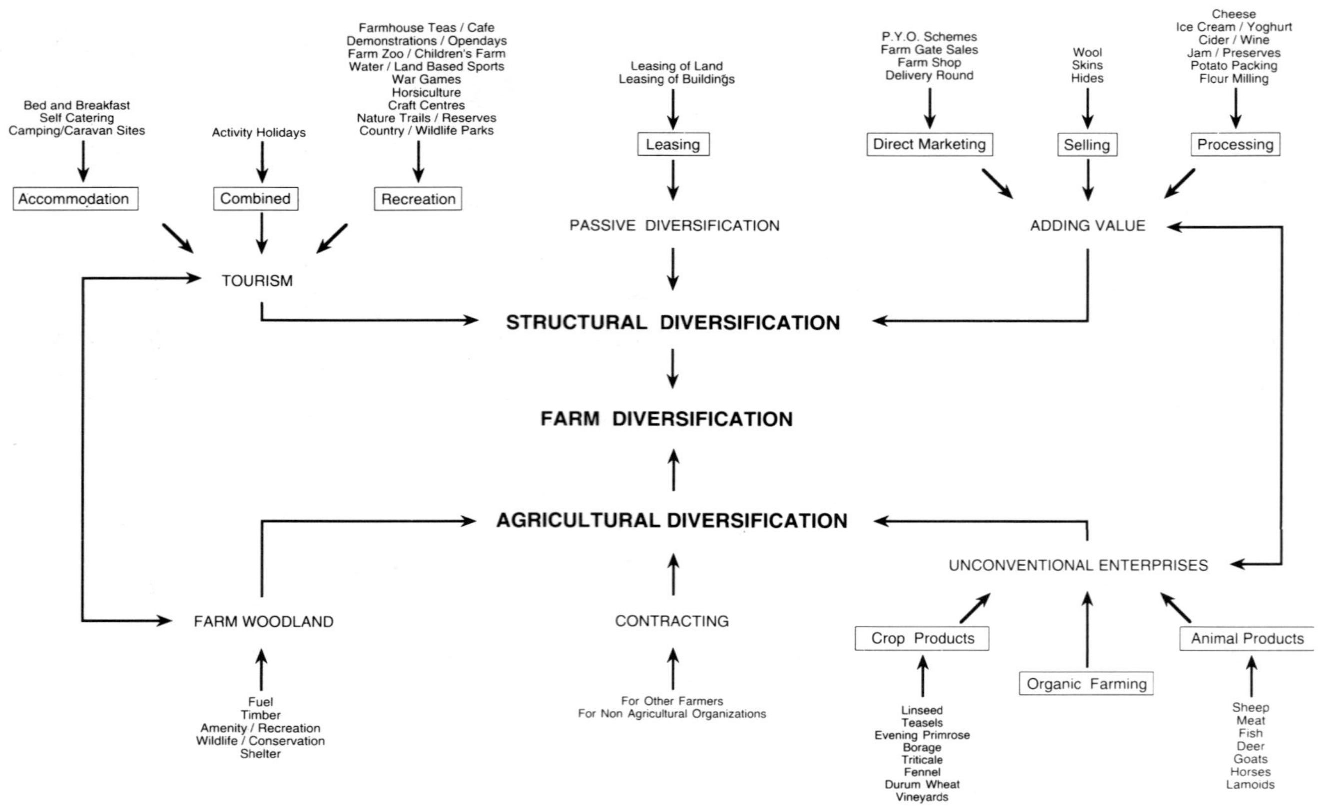

Fig. 7.1 A schematic flow diagram of agricultural and structural forms of farm diversification.

that are of relevance to farm diversification, namely the Farm Diversification Grant Scheme (FDGS) and the Farm Woodland Scheme (FWS).

The Farm Diversification Grant Scheme (FDGS)

The FDGS is described by MAFF as a comprehensive package of assistance to farmers considering diversifying into non-agricultural profit-making activities on their farms. It can be viewed as an attempt to maintain farmers' incomes while support prices for agricultural products are reduced and further production control measures are introduced (Bowler and Ilbery, 1992). The FDGS is related to farm-centred diversification and not pluriactivity; it is also restricted to structural types of diversification.

Eligible enterprises for financial support from the FDGS include: farm-based accommodation (although this category was removed in January 1991); educational ventures and various recreational activities; livery for horses and ponies; facilities for the hire of horses and ponies in Less Favoured Areas (LFAs); the processing of timber products; and the direct marketing and/or processing of agricultural produce.

Three types of grant are available under the FDGS.

1. *Capital grant*, available to assist the introduction or expansion of on-farm profit-making businesses of a non-agricultural kind. The rate of grant is 25% on investments between £750 (minimum) and £35 000 (maximum).
2. *Feasibility study grant*, of 50% towards the cost of a diversification feasibility study, up to a maximum of £3000 for individuals and £10 000 for a group.
3. *Marketing cost grant*, towards the cost of employing agents or separate personnel to conduct marketing and promotional functions such as the design and printing of brochures and the paying of promotional body fees. The grant is paid in each of three consecutive years, at decreasing rates of 40%, 30% and 20% up to a maximum of £3000 for individuals and £10 000 for a group.

Farmers can apply for any combination of the three grants; they can even have more than one capital grant, on the same or different schemes, up to the maximum of £35 000. However, farmers have to satisfy a number of criteria in order to qualify. First, they have to demonstrate the likely viability of the new enterprise and show that they could not carry out the investment programme without the grant. Second, they should derive more than half their annual income from both the new and existing farm enterprises and, third, spend at least 1100 hours per year working on the agricultural activities on the farm. Finally, they must either have been in farming for at least 5 years or hold a suitable training certificate.

The Farm Woodland Scheme (FWS)

Voluntary in nature, the FWS aims to encourage farmers to plant new woodlands on land currently in agricultural use; it is, therefore, related to agricultural and not structural diversification. More specifically, the FWS seeks to:

1. divert land from agricultural production and thus help reduce agricultural surpluses;
2. enhance the landscape, create new wildlife habitats, encourage recreational use and expand tourist interests;
3. contribute to supporting farm incomes and rural employment;
4. encourage greater interest in timber production from farms and, in the longer term, to contribute to the UK's timber requirements.

The FWS provides annual payments to farmers in addition to, but conditional upon receipt of, a planting grant from the Forestry Commission. It is confined mainly to arable land or improved grassland that has been cultivated and reseeded within the last 10 years. However, there is an allocation of 3000 ha (out of a total of 36 000 ha) for planting on unimproved land in LFAs. The rates of grant for both the initial planting and the annual payments are detailed in Table 7.1, which highlights the geographical differences between both areas of improved and unimproved grassland and location in or outside LFAs. The number of annual payments made to a farmer is dependent upon the species of trees planted. For oak and beech, payments are made for 40 years; this falls to 30 years for mixed areas containing more than 50% broadleaved species, 20 years for mainly coniferous areas, and 10 years for coppice.

To qualify for the FWS, a farmer must plant a minimum of 3 ha but no more than 40 ha of trees; each block of woodland has to be at least 1 ha. The farmer must be the occupier of the land in question and running an agricultural business; planting of a whole holding is, therefore, prohibited (because the agricultural business would disappear). There are other planting, management and operational rules that must also be met before an application is approved.

The Internal (Demand) Environment

Data are now available on the uptake of the FDGS and FWS during the first 2 years of their operation. In both cases, the response has been disappointing. The number of applicants for the FDGS between 1988 and 1990 amounted to only 1% of all farmers; the scheme has been biased towards southwestern and southeastern England (Fig. 7.2a) and dominated by accommodation and recreation enterprises (Ilbery and Stiell, 1991).

Table 7.1. Planting grants and payments under the Farm Woodland Scheme.

(a) Forestry Commission Planting Grants (£/ha)

Approved area (ha)	Conifers	Broadleaves
1 – 2.9	505	1375
3 – 9.9	420	1175
10 and over	240	975

(b) Annual payments under the Farm Woodland Scheme (£/ha)

	Less Favoured Areas		Elsewhere
	Severely disadvantaged	Disadvantaged	
Arable land and improved grassland[a]	100	150	190
Unimproved grassland	30	30	–

[a] Cultivated and reseeded in the last 10 years.

Although over 70% of applications have been approved, less than 50% of all approvals have been completed. In addition, the vast majority of approvals (over 90%) have been for the capital grant; very little interest has been shown in feasibility study and marketing costs grants.

The approved uptake of the FWS stood at just over 9000 ha after 2 years, which is well short of the 24 000 ha target for that time period. More worryingly, the application rate has been almost continuously downward since the initial flourish in October and November, 1988. As with the FDGS, there is a distinct southern bias in the pattern of adoption (Fig. 7.2b), with a higher than expected uptake in a band of counties just south of a line running from the Severn to the Wash.

Although the poor rate of uptake of both schemes can in part be related to inadequate promotion and marketing in the external environment, a number of 'resistance' factors are operating within the internal environment of individual farms. Consequently, the rest of this chapter will concentrate on the results of two surveys, one national and one local, which have attempted to analyse the attitudes of farmers towards the FDGS and FWS. Two particular aspects are of interest: first, the major farm and farmer characteristics that distinguish 'adopters' from 'non-adopters'; and second, the main reasons for adopting/not adopting the two schemes (as given by the farmers). Only brief details can be provided; further analysis on the FDGS can be found elsewhere (Bowler and Ilbery, 1992).

The national survey of the FDGS included 133 adopters and 370 non-

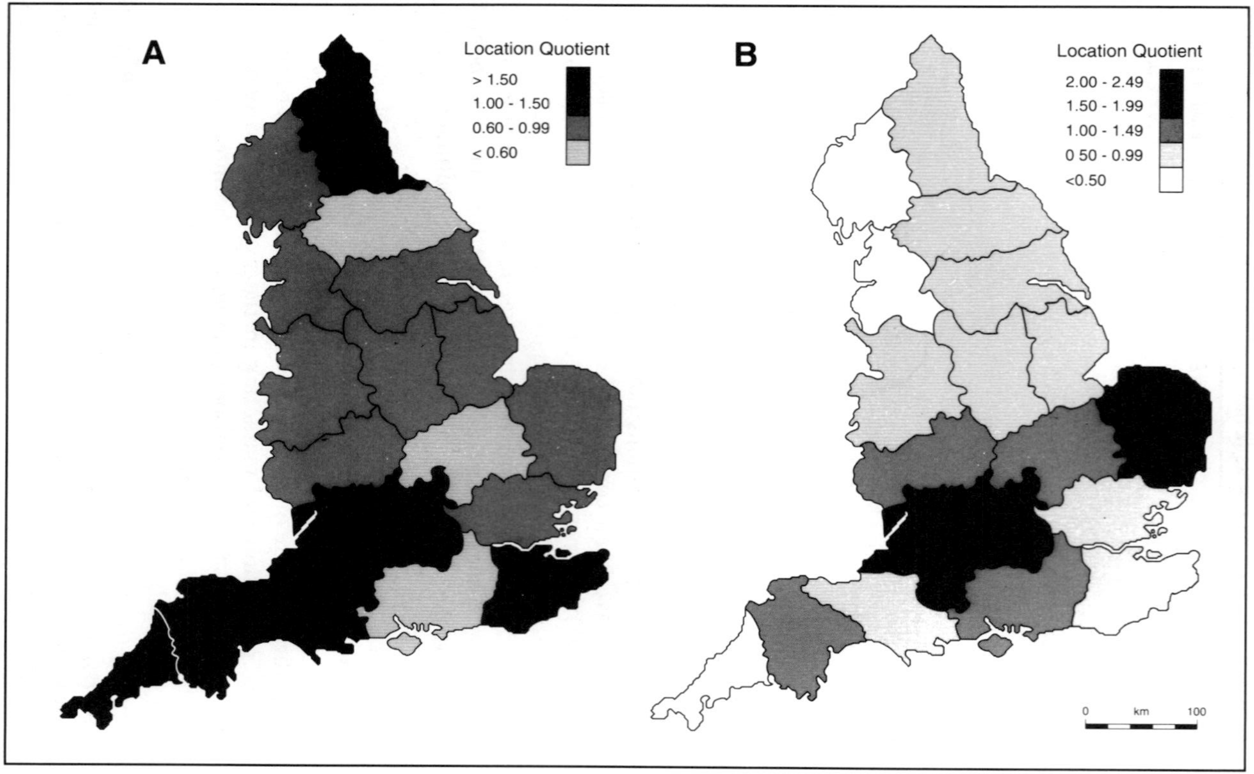

Fig. 7.2. Uptake of MAFF grants for diversification in England. A: concentration of approved capital grants under the FDGS; B: concentration of approvals under the FWS.

adopters, representative of the farm size, farm type and locational characteristics of all farms in England and Wales (Bowler and Ilbery, 1992). In contrast, the local pilot survey of the FWS involved interviews with 31 adopters and 37 non-adopters from 32 parishes in northeast Gloucestershire, an area with a higher than national rate of FWS uptake (Fig. 7.2b).

Farm and farmer characteristics

A number of consistent trends emerge when the characteristics of adopters and non-adopters are examined. Various farm characteristics are quite important in distinguishing adopters from non-adopters, notably farm type, farm size, net farm income and the level of indebtedness. A higher proportion of participants in the FDGS are of the cereal/cropping and sheep/beef farm types compared with non-participants; this situation is reversed for dairy, pigs/poultry and horticultural farm types. Such a contrast is not apparent in the FWS, but interestingly a much higher proportion of adopters (61%) than non-adopters (35%) have a tradition of woodland on their farms.

Adopters of both the FDGS and FWS tend to have larger farms than non-adopters, confirming the well-documented trend for larger farm businesses to be innovative in the adoption of grant-aided investment schemes (Bowler, 1976). Nearly 40% of FDGS adopters are over 40 BSU (Business Size Units), whereas only 26% of non-adopters are of that size; a similar difference exists in the FWS survey results (Table 7.2). In turn, adopters also tend towards higher net farm incomes than non-adopters; this is more marked in the FWS sample than in the FDGS survey. All the interviewed participants in the FWS have an annual income in excess of £10 000, with over 50% earning more than £20 000; in contrast, 40% of non-participants recorded net farm incomes of less than £10 000 (Table 7.2). Finally, adopters are having to finance a higher level of borrowed capital compared with non-adopters; it is not clear, however, whether this higher level of borrowing is a cause or effect of their diversification. Significantly, 42% and 38% of FDGS and FWS non-adopters respectively reported no borrowing at all. This helps to explain why so many non-participant farmers have not needed to adopt the two schemes; they have been under less financial pressure than those who are servicing large debts at a time of high interest rates.

With regard to farmer characteristics, adopters demonstrate many of the social features of innovators (Hagerstrand, 1967; Brown, 1981). For both schemes, adopters tend to be younger, have continued their full-time education after school, and have received more formal agricultural training (Table 7.3). All these features predispose adopters towards innovative behaviour. Age in particular appears to be an important influence on a farmer's decision as to whether or not to participate in the FDGS and the

Table 7.2. Farm characteristics of FDGS and FWS adopters and non-adopters (% in each group).

Farm size

FDGS (BSU)

	4–15.9	16–40.0	Over 40
Adopters	29	32	39
Non-adopters	33	41	26

FWS (ha)

	<50	50–99	100–200	>200
Adopters	10	6	55	29
Non-adopters	27	30	22	21

Annual net farm income (£000s)	<5	5–10	10–15	15–20	>20
FDGS					
Adopters	15	24	27	11	23
Non-adopters	42	21	15	7	15
FWS					
Adopters	0	0	3	45	52
Non-adopters	8	32	19	24	16

Capital borrowed (£000s)	None	<20	20–50	50–80	>80
FGDS					
Adopters	15	24	27	11	23
Non-adopters	42	21	15	7	15
FWS					
Adopters	10	22	42	16	10
Non-adopters	38	8	14	13	27

FDGS: Farm Diversification Grant Scheme; FWS: Farm Woodland Scheme
Source: Bowler and Ilbery (1992).

FWS. This is especially noticeable in the FWS, where over 60% of adopters are under 45 years of age in comparison with just 16% of the non-adopters. Older farmers tend to have left school at a relatively earlier age and not received as much agricultural training as younger farmers; as a consequence they have a more traditional attitude towards agricultural production and thus resist non-agricultural developments on their farms.

Over one-third of FDGS adopters stated that their spouse was either active in decision-making on the farm, or filled more than one role in running the farm business; the comparable figure for non-adopters was

Table 7.3. Farmer characteristics of FDGS and FWS adopters and non-adopters (% in each group).

Age (years)	<25	26–45	46–65	>65
FDGS				
Adopters	2	44	50	4
Non-adopters	1	32	57	10
FWS				
Adopters	3	58	39	0
Non-adopters	0	16	71	13

Full-time education (age)	<16	17–18	>18
Adopters	43	22	36
Non-adopters	69	16	15
FWS			
Adopters	13	35	52
Non-adopters	32	38	30

Agricultural training	Yes	No
FDGS		
Adopters	47	53
Non-adopters	26	74
FWS		
Adopters	55	45
Non-adopters	40	60

FDGS, FWS: see Table 7.2.
Source: Bowler and Ilbery (1992).

just 14%. The relationships between the farmer, the spouse and the farm business are important in the development of farm diversification. Where the spouse has a greater involvement, there appears to be a more favourable disposition towards farm diversification (Evans and Ilbery, 1989; Symes, 1991) and the FDGS itself. A further stimulus in this context is that a greater proportion of adopters have children wishing to continue the farm business. However, this was not substantiated in the case of farm woodland, where the FWS seemed to be particularly attractive to younger, less traditional farmers with a positive attitude towards conservation (see below).

Reasons for adoption/non-adoption

Certain indirect explanations of why farmers have responded differently to the incentives contained in the FDGS and FWS have already been highlighted. A more direct approach is to elicit reasons from the farmers themselves as to why they have/have not participated in the two schemes. The primary motive for adopting a capital grant under the FDGS was the generation of income by widening the enterprise base of the business (34% of replies). This was followed by the use of spare resources on the farm (19%) and benefiting from the farm's location (19%). An identical rank order of reasons was found in a study of farm diversification on the urban fringe of the West Midlands (Ilbery, 1988). All other reasons were relatively insignificant, although a number of farmers cited 'to offer people an opportunity to share the countryside' as an important motive. It is significant that nearly half the participants had used the FDGS to expand an existing diversified enterprise rather than to start new ventures. A *selectivity effect* appears to exist, whereby the grant is subsidizing farmers in carrying out development schemes they would have completed anyway (Gasson and Potter, 1988; Ilbery and Stiell, 1991).

Such a selectivity effect was also apparent among adopters of the FWS, because the joint primary reasons given were 'to finance a scheme that was already under consideration' and 'as a means of achieving conservation and enhancing the appearance of the farm'. It should be noted, however, that farmers' attitudes to conservation are not necessarily a good guide to their conservation behaviour (Carr and Tait, 1991). Indeed, of the 31 farmers who viewed conservation as either important or very important, only 17 had carried out conservation schemes on the farm, involving the planting and management of trees, in the 10 years before the FWS. The second group of reasons for adopting the FWS included the need to diversify and generate income, but also the desire to pursue a hobby. Generating employment on the farm and reducing agricultural productivity were not important reasons for participating in the FWS.

Among the factors given for not participating in the FDGS, a basic lack of interest in farm diversification emerged as the most important. Farmers attached great importance to continuing their employment principally as agricultural producers, without diversification. The latter is perceived to represent a surrender of a self-image as agricultural producers. Not wishing to borrow capital, or increase the present level of indebtedness, was given as the next important reason for non-participation. At a time of economic recession and high interest rates, borrowing capital to develop non-agricultural enterprises (with which farmers have little experience) is not necessarily a rational business decision. A third reason for non-adoption was the wish not to employ additional labour; if present farm-family labour resources are not available to run a diversified enterprise, farmers seem disinclined to employ hired labour. Significantly, there was a lack of criticism about the FDGS itself; the reasons for non-adoption lie in farmers' attitudes to diversification as a business strategy.

Three groups of factors were highlighted by the non-adopters of the FWS. The first group included the small size of grant, the lack of financial return in the short term, and insufficient knowledge of the scheme. These were emphasized by the older farmers, with low incomes and no family succession. A second group of factors related to capital availability and a lack of relevant skills and experience in woodland management. Many non-adopters were already in financial debt and thought they would have problems in obtaining sufficient capital. A combination of good quality land, lack of labour, reduction in farm flexibility and a conflict with tradition emerged as a third set of reasons for non-adoption. Once again, the regulations surrounding the FWS were not perceived as an important resistance factor. Non-adoption of the FWS does not, therefore, reflect a lack of interest, as in the FDGS, but rather a lack of relevant woodland skills.

Assessment and Conclusion

Farm diversification is not a new phenomenon. However, to farmers the idea of generating income from non-traditional 'alternative' enterprises is relatively new. Two sets of factors will interact to affect the adoption of farm diversification: first, the characteristics of farm diversification itself, including such features in the *external* environment as financial incentives under the FDGS and FWS; and second, the characteristics of farms and farmers in the *internal* environment.

Reasons for the slow rate of adoption of the two schemes are not difficult to find. Farm diversification is only one possible adjustment strategy and many farmers prefer other options. The national survey demonstrated that these include selecting alternatives more directly connected with traditional agricultural production, developing income through OGAs off the farm, and – significantly – doing nothing. The latter group tended to comprise farmers nearing retirement, with little/no debt and no heirs wishing to continue the business. A large number of farmers are just not interested in farm diversification; they have yet to be convinced of the concept.

A strong selectivity effect can be identified in the adoption of both the FDGS and FWS. Nearly half of FDGS adopters were using the capital grant to expand already existing alternative enterprises, just as adopters of the FWS have a tradition of farm woodland and openly stated that they were using the scheme to finance a project already under consideration. Adopters have a set of farm and personal characteristics that distinguish them from non-adopters, including larger farms, higher incomes, a greater willingness to borrow capital, younger in age, higher educational attainment, formal agricultural training and, especially in the case of the FDGS, children wishing to continue the farm business and a spouse more actively involved in decision-making on the farm.

Non-adopters appear to have a psychological resistance to the concept of diversification. Until farmers move from 'awareness' into 'interest' and 'evaluation' stages, adoption will remain relatively low. A need exists, therefore, to raise farmers' awareness and knowledge of what is actually available to them. This suggests that much more research is required on the ways in which public and private institutions in the external environment publicize policy measures and inform farmers. Indeed, further research is required on various dimensions of the external (supply) environment and their interactions with the internal (demand) farm environment.

Acknowledgement

The author wishes to acknowledge the financial support given by the Ministry of Agriculture, Fisheries and Food for an evaluation of the FDGS.

References

Baldock, D., Cox, G., Lowe, P. and Winter, M. (1990) Environmentally Sensitive Areas: incrementalism or reform? *Journal of Rural Studies* 6, 143–62.

Bowler, I. (1976) Adoption of grant aid in agriculture. *Transactions of the Institute of British Geographers* 1, 143–58.

Bowler, I. (1987) The geography of agriculture under the CAP. *Progress in Human Geography* 11, 24–40.

Bowler, I.R. and Ilbery, B.W. (1992) *Farm diversification in England and Wales – a comparison of participants and non-participants in the Farm Diversification Grant Scheme*. Occasional Paper 21, Department of Geography, University of Leicester.

Brotherton, I. (1991) What limits participation in ESAs? *Journal of Environmental Management* 32, 241–9.

Brown, L. (1981) *Innovation Diffusion: a New Perspective*. Methuen, London.

Carr, S. and Tait, J. (1991) Farmers' attitudes to conservation. *Built Environment* 16, 218–31.

Duncan, J. and Ley, D. (1982) Structural Marxism and human geography: a critical assessment. *Annals of the Association of American Geographers* 72, 30–59.

Evans, N. and Ilbery, B. (1989) A conceptual framework for investigating farm-based accommodation and tourism in Britain. *Journal of Rural Studies* 5, 257–66.

Fuller, A. (1990) From part-time farming to pluriactivity: a decade of change in rural Europe. *Journal of Rural Studies* 6, 361–71.

Gasson, R. (1988) Farm diversification and rural development. *Journal of Agricultural Economics* 39, 175–82.

Gasson, R. and Potter, C. (1988) Conservation through land diversion: a survey of farmers' attitudes. *Journal of Agricultural Economics* 39, 340–51.

Gasson, R., Crow, G., Errington, A., Hutson, J., Marsden, T. and Winter, M. (1988) The farm as a family business: a review. *Journal of Agricultural Economics* 39, 1–41.

Hagerstrand, T. (1967) *Innovation Diffusion as a Spatial Process*. University of Chicago Press, Chicago.

Haines, M. and Davies, R. (1987) *Diversifying the Farm Business*. BSP Professional Books, Oxford.

Harrison, A. (1975) *Farmers and Farm Businesses in England*. Miscellaneous Study No. 62, Department of Agricultural Economics and Management, University of Reading.

Ilbery, B. (1988) Farm diversification and the restructuring of agriculture. *Outlook on Agriculture* 17, 35–9.

Ilbery, B. (1991) Farm diversification as an adjustment strategy on the urban fringe of the West Midlands. *Journal of Rural Studies* 7, 207–18.

Ilbery, B. and Stiell, B. (1991) Uptake of the Farm Diversification Grant Scheme in England. *Geography* 76, 259–63.

Marsden, T., Whatmore, S., Munton, R. and Little, J. (1986a) The restructuring process and economic centrality in capitalist agriculture. *Journal of Rural Studies* 2, 271–80.

Marsden, T., Whatmore, S., Munton R. and Little, J. (1986b) Towards a political economy of capitalist agriculture: a British perspective. *International Journal of Urban and Regional Research* 10, 498–520.

Marsden, T., Whatmore, S. and Munton, R. (1987) Uneven development and the restructuring of British agriculture: a preliminary exploration. *Journal of Rural Studies* 3, 297–308.

Marsden, T., Whatmore, S., Little, J. and Munton, R. (1989) Strategies for coping in capitalist agriculture: an investigation of the responses of farm families in British agriculture. *Geoforum* 20, 1–14.

McInerney, J., Turner, M. and Hollingham, M. (1989) *Diversification in the use of farm resources*. Report 232, Department of Agricultural Economics, University of Exeter.

Munton, R. (1990) Farm families in upland Britain: options, strategies and futures. Paper presented to Association of American Geographers, Toronto, April.

Robinson, G. (1991) EC agricultural policy and the environment: land use implications in the UK. *Land Use Policy* 8, 95–107.

Rogers, E. (1962) *Diffusion of Innovations*. MacMillan, New York.

Shucksmith, D., Bryden, J., Rosenthall, P., Short, C. and Winter, M. (1989) Pluriactivity, farm structures and rural change. *Journal of Agricultural Economics* 40, 345–60.

Slee, R.W. (1987) *Alternative Farm Enterprises*. Farming Press, Ipswich.

Symes, D. (1991) Changing gender roles in productionist and post-productionist capitalist agriculture. *Journal of Rural Studies* 7, 85–90.

Unwin, T. (1988) The propagation of agrarian change in north-west Portugal. *Journal of Rural Studies* 4, 223–38.

Whatmore, S. (1991) Life cycle or patriarchy? Gender divisions in family farming. *Journal of Rural Studies* 7, 71–6.

Whatmore, S., Munton, R., Marsden, T. and Little, J. (1987) Interpreting a relational typology of farm business in southern England. *Sociologia Ruralis* 27, 103–22.

III

Changing Relationships Between Agriculture and the Environment

8

Nature, Uneven Development and the Agricultural Landscape

Rebecca Roberts

In much recent literature the importance of nature to the agricultural landscape tends to slip from view. Munton (1987, p. 31) for example, claims 'there is much greater interest in the social and economic processes that underlie change and much less in seeking to attribute changes in activity and appearance to the influences of location or physical environment'. And in their provocative analysis, Goodman *et al.* (1987) recognize the uniqueness that a natural production process imparts to agriculture but then highlight the substitution of industrial for natural processes; nature becomes a residuum. This shift in emphasis, of course, reflects larger trends in the social sciences where interest in exploring the implications of the dynamic social structures of capitalism is high. Rural studies are no exception; attention centres on social processes of change. The natural component of agriculture is, of course, recognized as essential but is treated as uninteresting or irrelevant to restructuring.

This chapter contributes to the effort to recover the link between social theory and nature. It asks the question: what is the role of nature in the restructuring of the agricultural landscape? It bases the analysis in the necessarily dialectical character of the nature–society relationship. The dialectic implies that not only do the social and the natural interact, but that prior interaction forms the causal preconditions of future action. The chapter first reviews efforts to define this relationship in the social sciences. It then uses the concept of uneven development to elaborate a geographic approach to the nature–society relationship that emphasizes the spatiality of that relationship. The analysis is centred in the social construction of natural resources; uneven development describes how social construction under capitalism is organized dynamically in space and time. Finally, historical–geographic analysis of the development of groundwater management institutions in the United States Southern Plains illustrates how uneven development can provide a linkage between nature and the agricultural landscape; this linkage avoids both environmental and social determinism to clarify the political economy of agricultural restructuring.

The Problem of Mid-level Theory

Given the recognition of the centrality of nature to agricultural production, the problem inhibiting the integration of both nature and society through dialectical analysis is the lack of mid-level theoretical concepts. On the one hand, the large scale at which capitalist forces are formed is difficult to integrate with the local specificity of human–nature interactions. On the other hand, an approach capable of incorporating the variability and specificity of these interactions tends to lose the larger social processes.

Attempts to incorporate nature in environmental sociology have generated considerable interest but have tended to founder on a simplified theoretical basis that leaves the way open to charges of environmental determinism (Buttel *et al.*, 1990). What is remarkable about the sociological debates as a whole is the prevalence of a non-dialectical, ahistorical, even atheoretical approach to both empirical and theoretical nature–society analysis. Determinist arguments – both environmental and social – continue to emerge. Environmental historians have resolved some of the polarities between simple environmental and social determinist positions. Reasons lie in the compatibility of the historical perspective with a dialectical logic and in the more careful interweaving of political economic theory into explanations. For example, two studies of transformation in colonial New England clarify the dialectic between cultural and ecological revolutions (Cronon, 1983; Merchant, 1989). But the analysis remains temporal; it promotes causality organized through time over causality organized through space (Soja, 1989). Explanation does not extend to the spatial organization of the New England landscape and its effects on further development.

Regional political ecology represents a very influential effort to increase the complexity and the spatiality of social analysis of the nature-society interaction (Blaikie and Brookfield, 1987). It brings natural and social forces together by highlighting the conjuncture between the natural and social inherent in the individual resource manager making decisions about resource use in a specific place. Resource degradation

> at one place and time will be conjunctural and complex. There are patterns that repeat themselves in human–environment relations, but this modelling can only be partial at best. Case-study material therefore becomes crucial, . . . an approach is suggested which allows for complexity, uncertainty and great variety, and one which takes as its point of entry those data which are beset with least

> uncertainty – the direct relationship between the land-user and manager and the land itself.
>
> (Blaikie and Brookfield, 1987, p. 16)

The approach successfully emphasizes the interactive and contradictory nature of the nature–society relationship. Case studies detail complex chains of causation that connect a richly detailed natural environment with social constraints and opportunities. At the same time, regional political ecology does not contain the conceptual tools to provide an equally rich insight into the political-economic processes that generate, and are generated by, the local social context (Watts, 1990). Social forces remain ultimately exogenous and mechanical because the scale of the social analysis proceeds from the bottom up.

Uneven development (Smith, 1984) provides a mid-level theoretical construct; it is capable of contributing insight into both the larger-scale social forces that structure the resource landscape and the way that those forces interact with nature to produce a mosaic of places. The concept ties the social process of production/reproduction dialectically to the landscape. In urban and industrial geography, where most of the applications of uneven development are located, fixed capital investment is the means by which social processes of capital are immobilized in the landscape. In the case of agriculture, the embedding of social processes is not so obvious. Yet the perceptible resource landscape arises from the interaction of natural and social processes that come together in technologies of agricultural production.

Uneven Development and the Agricultural Landscape

The social construction of natural resources

Resources are not simply physical substances but instead reflect a dialectic between society and nature. One pole of this dialectic is a social appraisal process deriving from the efforts of human beings to produce and reproduce their livelihoods. Appraisal reflects technology, investment in physical capital, transportation, markets, values and other institutions. Substances become resources when, as a result of these processes, they become useful. The other pole is nature and the possibilities and constraints inherent in its composition, arrangement and powers. The philosophical position is distinguished from earlier approaches, such as possibilism, by the mutual and historical causality inherent in the dialectic.

One of the most important implications of social construction is that natural resources are no longer static but become dynamic in a way that extends far beyond physical changes due to depletion or pollution. As technologies, capital investment, markets, or any number of factors change, what we define as natural resources also changes. Clearly, change in either the production or consumption processes restructures the nature–society relationship.

Some of the most interesting work comes from Marxist geographers (Burgess, 1978; Pepper, 1984; Smith, 1984; Redclift, 1987). There are important debates within this literature: on how the human part of nature relates to the non-human part, on whether historical materialism transcends the idealist–materialist dimension, over the importance of reproduction relative to production. Despite the apparently fundamental character of these controversies, differences may be less important than similarities once emphasis shifts to empirical analysis of concrete situations. All would agree that the nature–society relationship is intrinsically dialectical, historical and materialist. Further, they would all agree that the relationship arises in the basic labour process as people struggle to produce and reproduce their material existence.

Uneven development

The dynamic forces of capitalism impart patterns in time to the processes underlying the social construction of resources. Where technology is the means by which capitalists compete, the laws of motion result in what is described, in both neoclassical and Marxian analyses, as the technological treadmill. According to the technological treadmill concept, producers under intense competitive pressure adopt new technologies that reduce costs per unit of production. High profits accruing to early adopters are then gradually eroded as others follow suit and prices decline. But continuing competition ensures that the process will repeat itself with new innovations. Ultimately, competition through technological innovation becomes so general that the profits of innovation and early adoption, or relative surplus value, become the major source of profit (Harvey, 1982). A state of permanent revolution in the technologies of production has been reached.

Because a technology is a way in which people interact with nature in production, the treadmill has profound implications for the nature–society relationship. Permanent revolution in technology generates permanent restructuring in the nature–society relationship. The productive value of natural combinations in particular locations becomes fluid. Changing technologies may increase or decrease the relative value of particular natural characteristics. On the one hand, successive technologies tend to reduce the role of nature in production, especially in agriculture where nature represents a constraint limiting realization of profit (Goodman *et al.*, 1987). On the other hand, and paradoxically, these same technological innovations often tend to magnify previously unimportant distinctions in the qualities of land or other resources (Williams, 1980). This is especially evident where price competition following a rise in total production eliminates production from more marginal resources.

Uneven development describes how the technological treadmill and fixed investment are expressed through space to produce the landscape (Smith, 1984). It is clear that fixed capital investment is uneven in time and space, generating pattern to the built environment. It is also obvious that technological innovation is uneven in time. What is not so obvious is that technological innovation has an uneven expression in space as a result of the specificities of the production process and that this expression is correlated with natural features of the landscape. Further, the uneven spatial expression of technological innovations has the same consequences for the spatial pattern of agricultural development as does that of fixed capital for urban and industrial forms. The following discussion describes uneven development in terms of fixed capital and then generalizes it to take account of the role played by nature.

Concrete production processes take place in fixed locations where the tools and other conditions of production are brought together. Capital is mobile, seeking the highest profit, but only by fixing capital in the landscape can profit be realized through production. This tension between the fixity and mobility of capital generates uneven development. Investment in particular locations 'differentiates' production complexes, creating the conditions for localized production. But the same processes that differentiate also 'level'. Investments are superseded by new investments employing new technologies and taking advantage of new locations; the production advantages of old locations decline. These processes result in a generalized, simultaneous growth and decline, investment and disinvestment, differentiation and levelling, called uneven development.

In geography, uneven development has found applications mostly in urban and industrial development where fixed capital generates a spatial pattern to development; construction of a plant enables but also fixes an industrial activity. As a plant ages, it depreciates technologically and physically as newer, more modern facilities are constructed elsewhere. The similar causal role of nature in uneven development, however, has been underappreciated. Certainly the unevenness of localized depletion under capitalist production regimes is acknowledged, and geographers have noted the theoretical role of nature in uneven development (Duncan, 1989). But applications of uneven development in rural studies emphasize the differentiation of farm enterprises without incorporating the dialectic with nature (e.g. Marsden *et al.*, 1987, 1989).

Uneven development, however, is not just a product of capitalist social processes. Nature's forces are autonomous in the unfolding of uneven development, even if mediated by social forces of production. A technology is a means of combining labour and nature. The nature that enters this relationship is not homogeneous, as invariant physical mechanisms combine contingently in time and space. Not all the resulting combinations are equally well-suited for particular technologies. Hence, nature, as well as capital investment, fixes technologies into the variegated landscape to shape uneven development. Reinforcing sets of natural and social processes coalesce to create distinct ways of living and producing in particular places (Duncan, 1989). Capitalism, far from acting uniformly to homogenize places, also feeds the interaction that generates landscape differentiation (Williams, 1980).

What we have then, under uneven development, is a regional mosaic of differentiation and levelling, of increasing and decreasing productive advantage associated with natural features of the landscape. This implies not only a continual restructuring of the relationship between nature and production, but a geographic pattern to that restructuring. This does not imply that there is a mechanistic relationship between environment and production or that agricultural regions can be 'read off' from biophysical patterns. What it does mean is that technologies and other social processes dialectically help to determine which biophysical features will pattern production.

The political implications of uneven development are especially significant. Uneven development immerses individuals and localities in social and economic change, welcomed and unwelcomed. Individual enterprises and communities deploy the institutional means available to them to promote a bubble of growth or to preserve an island of stability against decline. These efforts, leading to place-based cross-class coalitions sometimes called growth coalitions, are structurally based in a confluence of interests, or 'structured coherence', derived from uneven development (Harvey, 1985). Much of the politics of natural resources, and much of the politics of place, reflect the efforts of such place-based coalitions, and factions within coalitions, to appropriate the gains and displace the losses of uneven development.

Uneven Development in the Southern High Plains

The example of groundwater development in the US Southern High Plains will clarify this conceptual argument (Roberts and Emel, 1992). At its heart it asserts that production and nature are dialectically related and that the concept of uneven development provides critical insight into this relationship and its geographic expression. As production is continually restructured, so will be the geography of natural resources.

The semiarid Southern High Plains of the Texas and New Mexico border overlie the Ogallala Aquifer, a massive aquifer cut off by erosion from its recharge area in the Rocky Mountains (Fig. 8.1). Recharge rarely

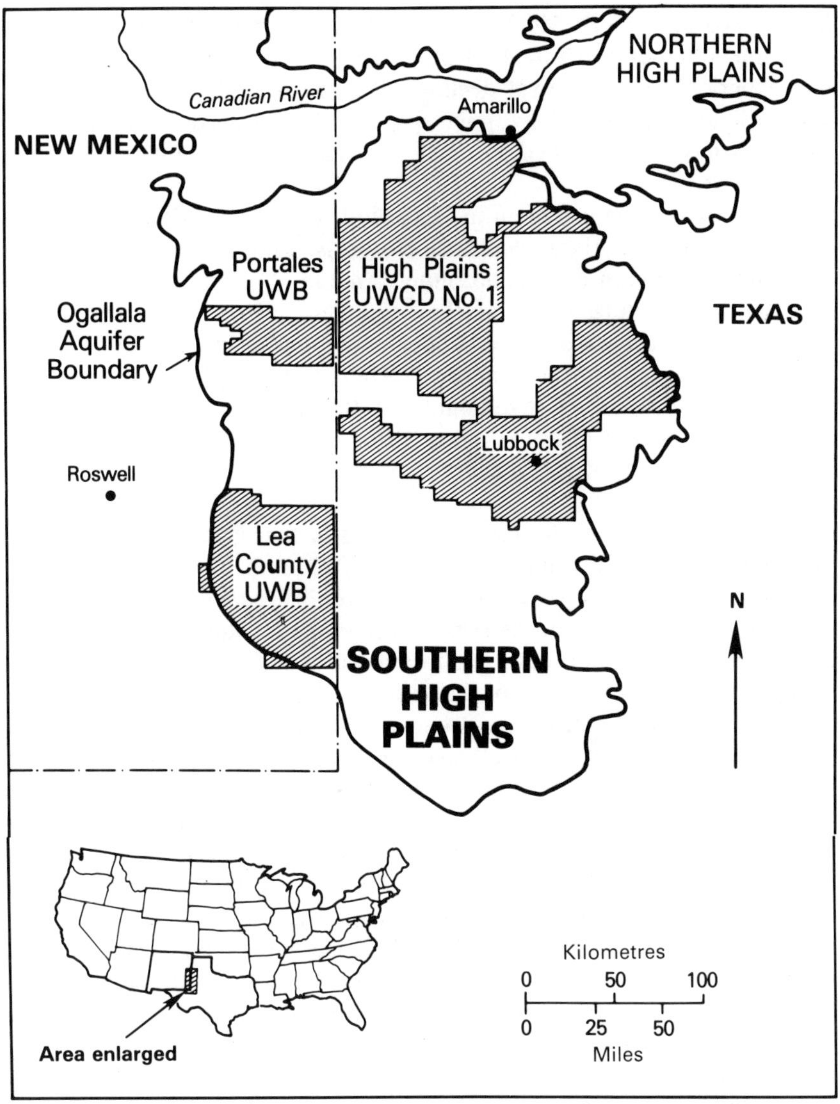

Fig. 8.1. Groundwater management districts in the Southern High Plains (Roberts and Emel, 1992). (Reprinted with permission from *Economic Geography* 68, 1992.)

exceeds a fraction of an inch, turning the aquifer, for all intents and purposes, into a basin of finite, fossil water. The aquifer remained undeveloped until a conjuncture in the 1930s of the Dust Bowl and improvements in pump technology made development desirable and technically feasible (Green, 1973). Irrigation burgeoned in the two decades following the

Second World War. By 1980, a quarter of the total aquifer area, comprising a much higher percentage of the effectively developable aquifer area, had experienced water-level declines of more than 25% (Luckey *et al.*, 1981). Significant areas have gone out of irrigated production, and most irrigators have been forced to make some accommodations to declining water levels.

Groundwater management in Texas and New Mexico developed under very different water rights doctrines (Roberts, 1992). Texas law relied on the English common law of 'absolute ownership', permitting landowners to withdraw as much water as they could use. To the extent that one pumper can affect water levels under neighbouring properties, this rule results in a 'tragedy of the commons' (Hardin, 1968). New Mexico, however, shifted early in its development to the doctrine of 'prior appropriation' – first in time, first in right. 'Prior appropriation' sets a quantity limit, as well as a priority, on every right, reducing the externality between pumpers.

When irrigation development stimulated further groundwater management legislation, Texas and New Mexico each took steps consistent with their distinct water doctrines (Roberts, 1992). Texas legislation permitted groundwater management districts to undertake a wide range of management actions, but made both the formation of a district and any actions that it might take voluntary, local decisions. The High Plains Underground Water Conservation District has become a very visible advocate of this approach; it regulates only well spacing and tail-water waste but actively encourages technologically efficient water use. New Mexico, on the other hand, instituted management districts formed and controlled from the central State Engineer's office. A management goal of two-thirds depletion over 40 years limits the quantity of water that can be appropriated and, therefore, protects pumpers from unplanned water-level declines. The fact that New Mexico specifically limits the external effects a pumper can impose on another has led to a widespread perception that New Mexico controls 'tragedy of the commons' problems much more effectively than Texas and, therefore, is a superior management system (Roberts and Emel, 1992).

The processes of uneven development are clearly evident in the history of groundwater development in the Southern High Plains. Differentiation and levelling, closely associated with the fortunes of the area, resulted from interactions between the variegated natural base and the social processes of production. The result has been a shifting geographic pattern of advantage for irrigated production. The development of turbine pump technology in the 1920s and 1930s made groundwater into a resource for large-scale irrigation (Green, 1973). Resource qualities were differentiated according to technological and hydrological characteristics that determined how much a well could pump; irrigation rapidly expanded after World War

II where water and soils were favourable. The development of sprinkler technology, particularly the centre pivot, subsequently expanded and shifted the range of suitable soils and slopes. Recently, water-conserving application systems have mitigated the effects of physical depletion.

Levelling proceeded simultaneously and, in recent years, has predominated. Physical depletion is, of course, an important leveller, but it would be a mistake to overemphasize its importance. Production often declines in value or stops long before physical depletion. Competition with increased production from other regions, advantaged by their own differentiations, has intensified the cost-price squeeze in the High Plains and led to the abandonment of marginal groundwater resources. In the 1950s and 1960s, feed grain production from the Midwest, advantaged by hybrid technology, and cotton in the Southwest, irrigated by subsidized water from federal water projects, dealt costly blows to High Plains producers. Similarly, increasing energy prices have disadvantaged pump irrigation relative to rain-fed and surface-water irrigated agricultural regions.

Both the dynamic character and spatiality of uneven development of the agricultural resource base are evident in these descriptions. In other words, the quality of the environment for production has been constantly reconstructed in an uneven geographic mosaic. Resource complexes are differentiated and then levelled, sometimes with very little change in physical characteristics. Technological or other social developments that differentiate in one locale serve under the competitive pressures of capitalism to level elsewhere. These changes in the value of the product of resource use figure prominently in the resource problems as defined by farmers and residents of these areas: the relative advantage and costs of rapidly changing irrigation technologies, the priorities of the agricultural research establishment, the cost of energy for pumping, and the declining profit rate on irrigated production (Walsh, 1980). Physical depletion, the focus of much analysis, is by no means the only, and not even the most important, source of this dynamism.

Structured coherences of interests deriving from uneven development underlie the sectoral and cross-class alliances that typify regional politics (Harvey, 1985). Growth coalitions serve as the focal point for the inevitable conflicts over who will appropriate the gains and suffer the losses of uneven development. In the Southern High Plains, where much of the economic surplus has been generated by irrigated agriculture, groundwater management institutions have figured prominently in regional politics (Roberts and Emel, 1992). But their role has not been that traditionally accorded to groundwater management in neoclassical theory. In the years immediately after the Second World War, local elites faced similar material conditions on both sides of the Texas–New Mexico border: new pump technology waiting to be exploited, similar favourable economic conditions for irrigated agriculture, similar soils and climate, and an available groundwater supply of varying quality. Yet, in efforts to appropriate the benefits, they faced very different political and legal environments.

The immediate problem, in both Texas and New Mexico, was to ensure access rights to groundwater for local farmers. In both cases, farmers were joined by local town and city elites: by bankers, by farm supply and processing interests, and by town and city commercial interests. In Texas, the major political task was to protect farmers' free access to groundwater; such access was threatened by the doctrine of 'absolute ownership' from East Texas urban and industrial interests seeking to ensure their long-term access to water supplies through a shift in water-law doctrine to 'prior appropriation' (Green, 1973). The High Plains growth coalition mobilized and was able, after a battle over many years, to fend off the threat. They substituted a law that enabled groundwater management but left it to voluntary action by voluntarily formed local management districts. This action preserved access to water for High Plains farmers and effectively pre-empted future state-controlled groundwater management.

In New Mexico, the problem presented by the new groundwater irrigation opportunities was to increase groundwater access by introducing flexibility into a managed prior appropriation water-rights system (Harris, 1954, 1958). 'Prior appropriation' had long been established as New Mexico's basic water doctrine and had been extended to groundwater in the 1920s. But 'prior appropriation' is predicated on a recurring flow of water; a water user receives a water right under the understanding that the flow to senior appropriators will not be affected in future years. The Ogallala, however, represents a finite stock of water; all appropriations after the first reduce future supplies to senior appropriators. Based on this logic, the State Engineer closed the Lea County basin to further development in 1948 at an early stage in its development. Lea County farmers united with local business interests to expand access. Test cases in the courts and negotiations with the State Engineer resulted in a new management goal that encouraged irrigation expansion by permitting a 66% depletion of the basin over 40 years. The relationship between High Plains irrigation interests and the State Engineer has been marked ever since by an accommodation useful to both.

In summary, political development of groundwater management institutions ensured secure and widely dispersed access to groundwater in both the Texas and New Mexico High Plains. Although the problems of agricultural production were similar, the form of the struggle, and the ultimate form of the management institutions, resulted from interplay between the similar interactions with nature in production, and the different juridical, political and cultural contexts in each state. These divergences, combined with the theoretical hegemony of the 'tragedy of the commons', were so great that they camouflaged the essential similarities in what was attempted and accomplished. Accumulating evidence indicates that the externality necessary to the development of the commons tragedy is small in this case (Roberts and Emel, 1992). Further evidence indicates that the New Mexico and Texas approaches are equally effective in controlling water-level declines through more efficient water use. Thus, the political expression of uneven development becomes a more informative model than the 'tragedy of the commons' for understanding the development of groundwater management institutions in the High Plains.

Conclusion

Restructuring of production generates a changing geography of agricultural development. This geography is related not only to capital investment patterns but to natural patterns because a production technology, by definition, combines nature and physical capital with a labour process. This implicates nature, fundamentally and dialectically, in uneven development under capitalism. Simultaneous forces of differentiation and levelling generate a changing agricultural landscape mosaic that reflects natural patterns and serves as a causal force in the development of a politics of place. Uneven development poses a dynamic resource environment to political coalitions structured by common interests in the fortunes of places. Resource conflicts routinely reflect efforts of such coalitions to possess gains and shift losses arising from uneven development.

References

Blaikie, P. and Brookfield, H. (1987) *Land Degradation and Society*. Methuen, London.

Burgess, R. (1978) The concept of nature in geography and Marxism. *Antipode* 10, 1–11.

Buttel, F.H., Larson, O.W. and Gillespie, G.W. (1990) *The Sociology of Agriculture*. Greenwood Press, New York.

Cronon, W. (1983) *Changes in the Land: Indians, Colonists, and the Ecology of New England*. Hill and Wang, New York.

Duncan, S. (1989) Uneven development and the difference that space makes. *Geoforum* 20, 131–9.

Goodman, D., Sorj, B. and Wilkinson, J. (1987) *From Farming to Biotechnology*. Basil Blackwell, London.

Green, D.E. (1973) *Land of the Underground Rain: Irrigation on the Texas High Plains, 1910–1979*. University of Texas Press, Austin.

Hardin, G. (1968) The tragedy of the commons. *Science* 162, 1243–8.

Harris, C.D. (1954) Groundwater law in New Mexico. *Journal of the American Water Works Association* 46, 10–18.

Harris, C.D. (1958) Water allocation under the appropriation doctrine in the Lea County Underground Basin of New Mexico. In: Haber, D. and Bergen, S.W. (eds), *The Law of Water Allocation in the Eastern United States*. The Ronald Press Company, New York, pp. 155–64.

Harvey, D. (1982) *The Limits to Capital*. Chicago University Press, Chicago.

Harvey, D. (1985) *The Urbanization of Capital*. The Johns Hopkins University Press, Baltimore.

Luckey, R.R., Gutentag, E.D. and Weeks, J.B. (1981) *Water-level and saturated-thickness changes, predevelopment to 1980, in the High Plains Aquifer in parts of Colorado, Kansas, Nebraska, New Mexico, Oklahoma, South Dakota, Texas, and Wyoming*. Hydrologic Investigations Atlas HA–652. US Geological Survey, Washington DC.

Marsden, T.K., Whatmore, S.J. and Munton, R.J.C. (1987) Uneven development and the restructuring process in British agriculture: a preliminary exploration. *Journal of Rural Studies* 3, 297–308.

Marsden, T.K., Munton, R.J.C., Whatmore, S.J. and Little, J.K. (1989) Strategies for coping in capitalist agriculture: an examination of the responses of farm families in British agriculture. *Geoforum* 20, 1–14.

Merchant, C. (1989) *Ecological Revolutions: Nature, Gender and Science in New England*. University of North Carolina Press, Chapel Hill.

Munton, R.J.C. (1987) Research in rural geography in Britain: some reflections on future directions. *Netherlands Geographical Studies* 27, 30–40.

Pepper, D. (1984) *The Roots of Modern Environmentalism*. Routledge, London.

Redclift, M. (1987) The production of nature and the reproduction of the species. *Antipode* 19, 222–30.

Roberts, R.S. (1992) Groundwater management institutions. In: Kromm, D. and White, S. (eds), *Groundwater in the High Plains*. University Press of Kansas, Lawrence, pp. 88–109.

Roberts, R.S. and Emel, J. (1992) Uneven development and the tragedy of the commons: competing images for nature-society analysis. *Economic Geography* 68 (in press).

Smith, N. (1984) *Uneven Development: Nature, Capital, and the Production of Space*. Basil Blackwell, Oxford.

Soja, E. (1989) *Postmodern Geographies: The Reassertion of Space in Critical Social Theory*. Verso, London.

Walsh, J. (1980) What to do when the well runs dry. *Science* 210, 754–6.

Watts, M. (1990) Review of *Land Degradation and Society*. *Capitalism, Nature, Socialism* 4, 123–31.

Williams, R. (1980) *Problems in Materialism and Culture*. Verso, London.

9

Agricultural Externalities and the Environment in the United States

Duane Nellis

Rural America consists of millions of hectares of land, in thousands of administrative jurisdictions, in a wide diversity of geographic regions, economic conditions and changing land-use patterns. The result of this variety of environmental conditions in America's rural areas – from remote wilderness to the rapidly expanding urban–rural fringe – is a high degree of programmatic idiosyncrasy. Given this diversity and breadth, the focus of this chapter is to synthesize the evolving relationship between agricultural externalities and the environment in the United States. The costs associated with environmental externalities in relation to agriculture are tremendous and need to be more fully recognized in sustainable agricultural systems.

Historically, rural areas in the United States have often been lax in protecting natural areas and enforcing environmental laws. In many instances such areas lack adequate public funds for monitoring and enforcement of environmental regulations (Daniels *et al.*, 1989). While sustainable economic development coupled with effective management is a worthy goal, depressed economies generally cause rural areas to place economic development efforts far above environmental protection priorities.

In the past two decades, however, United States congressionally mandated environmental land-use policy, together with pressure from various environmental groups, have created external pressures resulting in a framework for a major restructuring of America's agricultural system and associated rural land-use patterns. The National Environmental Policy Act, the Clean Water Act, the Food Security Act, and the Food, Agriculture, Conservation and Trade Act, in conjunction with state and local land-use planning regulations, have had a long-term influence on how lands will be used. Indeed, environmental issues are bringing many new players to an agricultural policy process that, historically, has included just farmers. Such groups as the Audubon Society, Nature Conservancy and the Sierra Club now have staff or members devoted to agricultural policy analysis. The new diversity of interests will make the difficult task of assessing the economic and social costs and benefits of environmental solutions even more complex (Drabenstott and Barkema, 1990). Despite these controls and interests, however, environmental problems associated with wetlands, groundwater depletion, soil erosion, reduction of shelterbelts, and conversion of prime agricultural lands to urban uses, for example, continue at alarming rates. This is further highlighted by the evidence that about half the farmland in the US has had a conservation plan at one time or another; but less than 8% of that land today is currently considered to have adequate conservation measures (Steiner, 1987).

Evolving Agricultural Externalities in Relation to the Environment

A consensus of opinion now appears to exist, even among the defenders of agricultural modernization, that the evolving complex of agricultural technologies and farm policies is creating damaging impacts on the environment. As a result, a major shift in attitudes is taking place towards environmental issues as they relate to agriculture and rural land use.

The agricultural sector's environmental externalities (costs borne by society and not reflected in market prices for agricultural commodities) have been recognized but largely unexamined or unquantified until recently (Cook, 1989). Such externalities are very large. A 1985 study by the Conservation Foundation, for example, estimated that sediment eroding from agricultural land into rivers, lakes, drainage ditches and other surface waters in the United States causes damages in the order of US$3 billion to US$13 billion annually (Clark and Haverkamp, 1985). A recent US Department of Agriculture study of groundwater contamination from agricultural chemicals estimated that the cost for potentially affected wells could range from US$0.9 to US$2.2 billion (Nielson and Lee, 1987). Others suggest that off-site costs of erosion in the US exceed the on-farm damages to productivity. Such externalities have gained considerable attention in various forms of environmentally related legislation that will be discussed later in this chapter.

There is also a growing farmer concern about a decline in groundwater quality. This represents a profound shift from post-war US farming attitudes (Cook, 1989). Public concerns about pesticide safety and their link to groundwater contamination have had a significant impact on pesticide use, including government bans or voluntary withdrawals of pesticide products. In addition, a large number of farms are questioning the routine use of agricultural chemicals and even the information provided by the agricultural chemical industry. A 1986 poll in Iowa, for example, found that 78% of Iowans, and over half of all farmers, favoured limits on chemicals to protect groundwater (Pins, 1986).

In the late 18th century, Edmund Burke wrote in *Reflections on the Revolution in France*, 'The most wonderful things are brought about, in many instances, by means, the most absurd and ridiculous'. The current widespread interest in sustainable agriculture (see Part V) would indicate that Burke may have been right. Who would have believed that a handful of ridiculous and absurd farmers, acting contrary to conventional wisdom by raising crops without chemical inputs and introducing diverse cropping systems that reduce their government payments, would start a revolution in agriculture? (Kirschenmann, 1991). For economic and environmental reasons, farmers across the United States are experimenting with a continuum of alternative agricultural systems. The National Research Council's findings on alternative agriculture lend support to the proposition that farming systems that reduce the use of agricultural chemicals and conserve natural resources can compete with conventional systems (Cook, 1989).

Another significant environmental externality influencing agricultural systems and sustainability is the shifting attitudes within Federal, State and local government, as well as agricultural research and extension. A small but growing number of agricultural researchers in the social and natural sciences are beginning to explore environmental aspects, in conjunction with economic and agronomic aspects, of alternative farming systems. The 1990 Farm Bill strengthens this focus by mandating a redirection of the US Department of Agriculture's research on sustainable agriculture. Priority will be placed on funding programmes that reduce the use of chemical pesticides, fertilizers and toxic natural materials (Cohen *et al.*, 1991). This is strongly reflected in new research agendas associated with agricultural research universities throughout the United States.

So often government policy has been put forth with little regard for environmental consequences. Within the soil conservation community, for example, it has long been recognized that commodity programme rules often frustrate conservation by rewarding the intensive production of specialized crops without regard for land capability. There are currently some 400 federal rural development programmes administered by 27 different US agencies. Not surprisingly, there has often been a lack of coordination between the various federal programmes, and some programmes actually work against each other. This tendency has both frustrated orderly rural development and, paradoxically, has actually increased the dependence of rural areas on the federal government (Daniels *et al.*, 1989). The new Farm Bill of 1990, however, provides greater opportunities for coordination. The Food, Agriculture, Conservation, and Trade Act (FACTA) of 1990, commonly called the '1990 Farm Bill', will further form the statutory foundation of US Agricultural Policy for much of the coming decade by strengthening the '1985 Farm Bill.' Of FACTA's 25 titles, 14 deal directly with some aspect of natural resource conservation, environmental protection or consumer issues. In addition, through the 1990 Farm Bill, an Agricultural Council on Environmental Quality was created to assist in developing a 5-year Environmental Quality Statement and Implementation Plan regarding agriculture and the environment (Cohen *et al.*, 1991).

Increased attention to agricultural issues within the environmental community is also dramatically impacting the structure of rural systems. Prior to the 1985 Farm Bill, the environmental community's involvement with federal policies affecting agriculture was confined to debates over federal law that regulates pesticides (Cook, 1989). Enactment of the Conservation Title of the 1985 Food Security Act marked what is likely to be a permanent involvement of environmental, conservation and wildlife groups with mainstream agricultural policies. Further involvement of these same groups in the development of the 1990 Farm Bill has strengthened the role of interest groups in environmental legislation.

The relationship between farm programme costs and environmental ethics is also changing. Up to September 1989, federal assistance to farmers in the United States, through the Commodity Credit Corporation (CCC), had exceeded US$174 billion dollars since the recession began in 1982. This is approximately US$87 500 per farm (of the approximately 2 million farms in the US). Approximately 24% of the US$11.8 billion in 1986, and US$16.7 billion in 1987, went to farms with sales of US$250 000 or more. These farms represent 4.6% of all US farms (Cook and Hinkle, 1989). In many years during the 1980s, the federal government spent as much money on corn (maize) farmers as it has spent on all Environmental Protection Agency programmes. The point here is that the government is beginning to demand a greater accountability for environmental protection for its agricultural dollars. In addition, farm credit organizations have often operated without regard to the environmental consequences of their lending. This attitude, however, is also changing.

Conflicts Between Rural Land Development and Preservation

Despite a greater conscientiousness in the US regarding the role of environmental externalities in agricultural systems, the past few decades have been a tale of conflict between development and preservation; between those who want to extract the maximum economic benefits from the land and those who struggle to conserve the US national wealth. And despite efforts through policy decisions by state and federal government to regulate environmental change, restructuring of the rural land-use system often continues at the expense of the environment, and without weighing environmental externalities. In addition, failure to distinguish between restructuring that affects the capacity of agriculture to produce food and fibre, and restructuring that affects the amenities associated with open space on rural residential life, has led to confusion in the effort to develop relevant farmland proposals. If the objective is to protect productive capability on the grounds that the state relies heavily on a strong agriculture and is a major supplier to food supplies, state involvement in farmland use in particular appears critical (Gardner and Wood, 1984).

Although less than half the United States' original 81 million ha of wetlands remain, an average of 20 ha disappear every hour of every day. The majority (87%) are converted to agriculture (Harger, 1990). Wetlands are among the most productive ecosystems in the world. Not only do 30% of the nation's endangered species depend on wetlands, but they also act as buffer zones against floods and drought, and help to purify water by trapping sediment. The Swampbuster provisions of the 1985 Farm Bill were unsuccessful for a number of reasons. First, the law only penalized farmers for draining wetlands if commodity crops, such as corn and wheat, were planted. A farmer was also free to drain a wetland to plant non-commodity crops, such as hay, and could still receive government price supports for commodity crops on other farmland. These problems, however, have been reduced through the 1990 Farm Bill (Cohen *et al.*, 1991).

Wind and water erosion on croplands in the US total 2.4 billion tonnes each year. This erosion can reduce the land's productivity by 60%. Sediment eroded from agricultural lands causes between US$3 and US$13 billion worth of damage to lakes and streams each year (Harger, 1990). Conservationists consider the passage of the soil conservation provisions of the 1985 Food Security Act (and strengthened in 1990) an environmental victory. About 18 million ha of highly erodible cropland is protected by the Conservation Reserve Program (CRP) (created by the 1985 law and extended in the 1990 Farm Bill). The programme is expected to reduce cropland erosion by 20%.

Based on estimates of these primary effects, the present value of net benefits of a 111 million ha CRP could range from US$3.4 to US$11.0 billion (Table 9.1) (Young and Osborn, 1990). These values could go up as newly eligible lands for the CRP include shelterbelts, windbreaks, permanent grass strips and waterways, stripcropped land and land deemed 'environmentally sensitive'.

Regional enrolment patterns suggest some limitations to the effectiveness of the CRP programme. Table 9.2 reveals the regional allocation of CRP participation. Two patterns are noteworthy. First, the western United States (i.e. the northern plains, southern plains, mountain and Pacific coast states) have a greater percentage of total CRP enrolment

Table 9.1. National income gains and losses from the Conservation Reserve Programme (CRP).

Category	Value (US$ Billion)
Gross income gains:	
Landowners	
Net farm income	9.2 – 20.3
Timber production	4.1 – 5.4
Natural resources/environment	
Soil productivity	0.8 – 2.4
Surface water quality	1.9 – 5.3
Filter strip water quality	0 – 0.3
Wind erosion	0.4 – 1.1
Wildlife	3.0 – 4.7
Gross income losses:	
Consumer costs	(12.7 – 25.2)
Establishing cover crops	
Landowner's share	(1.6)
Government's share	(1.6)
Technical assistance costs	(0.1)
Net programme benefit	3.4 – 11.0

Source: Young and Osborn (1990). (Extract reprinted with permission from *Journal of Soil and Water Conservation* 45, 1990.)

Table 9.2. Regional Conservation Reserve Programme (CRP) Enrolment.

Region	Region's cropland eligible[a] (%)	Total US cropland (%)	US total CRP accepted (%)
Northeast	17.69	4.34	0.43
Appalachian	21.59	6.92	3.31
Southeast	15.53	4.07	4.34
Delta states	11.97	3.73	2.95
Corn belt	17.94	23.58	14.25
Lake states	12.71	7.93	8.64
Northern plains	13.73	18.28	22.64
Southern plains	20.59	13.12	16.31
Mountain states	21.68	13.32	20.88
Pacific coast	14.61	4.71	6.27
US average or total	16.71	100	100

Source: Modified from Ervin (1988).

[a] Eligible cropland includes cropland in land capability classes II–V eroding in excess of three times the soil loss tolerance level, all cropland in land capability classes VII–VIII, or cropland with an erodibility index of 8 or greater. Total eligibility is limited to 25% of the total county cropland except in those cases where this limit has been waived by the Secretary of Agriculture. (Extract reprinted with permission from *Conservation Now* 1, 1989.)

relative to the sum of their eligible proportions. They account for nearly two-thirds of all the area enrolled, but only 49% of eligible hectares. In contrast, the Corn Belt and Appalachian states show the reverse enrolment pattern. While just over 30% of the eligible area resides in those two regions, only 17.5% of enrolments came from them (Ervin, 1988).

A companion programme to the CRP, nicknamed 'Sodbuster', terminates federal benefits if farmers plough highly erodible lands without an approved conservation plan. Farmers are breaking fragile lands at the same time as they are enrolling other lands in the CRP. Further, the erodible lands that have been preserved under the 1985 law are protected only temporarily. Farmers were paid up to US$50 000 dollars to take land out of production, but most of the leases expire in 10 years (Harger, 1990).

Groundwater contamination and abuse is another indicator of conflict between rural development and preservation. The Environmental Protection Agency (EPA) has found traces of 46 pesticides in the groundwater of 26 states (Harger, 1990). At least 18 of those detected chemicals are listed on EPA's list of pesticides that could be carcinogenic. Groundwater is the main source of drinking water for more than half the US population. Beyond demand for groundwater, chemical runoff threatens lakes and streams, killing fish and wildlife.

Groundwater demand for irrigation continues to increase. In areas overlying the Ogallala Aquifer, withdrawal rates far exceed recharge, yet incentives still exist for continuing to mine groundwater for short-term economic gain. In the Ogallala Aquifer region, saturated thickness levels vary from 0 to over 300 m, with variations of 0–200 m in Kansas (Fig. 9.1). Four-fifths or more of the Ogallala Aquifer water goes for irrigation. If conservation to extend the life of the aquifer is to have any meaning, it must occur in agriculture. Farmers are both the users of the water and the people most directly dependent on its continued availability: thus there is a built-in motivation to conserve. Fortunately, there are many practices, techniques and devices that can be used to improve irrigation efficiency (Kromm and White, 1990). In many parts of the Kansas Ogallala Aquifer area, however, groundwater supplies are predicted to run out by 2015 to 2020. This is despite implementation of institutional pressures associated with the passage of the Kansas Groundwater Management District Act, and despite farmers appearing to be more aware of the long-term advantages of changing to more water-efficient crops (Nellis, 1987).

Affecting much of the Ogallala region is a Buffalo Commons proposal based on the assumption that human interaction with the environment, as existing since European settlement, cannot continue (Dawson, 1988). The Poppers (proposers of this idea) believe that people will continue to leave the Great Plains due to groundwater depletion and economic decline until the region is almost completely depopulated. They suggest the purchase of

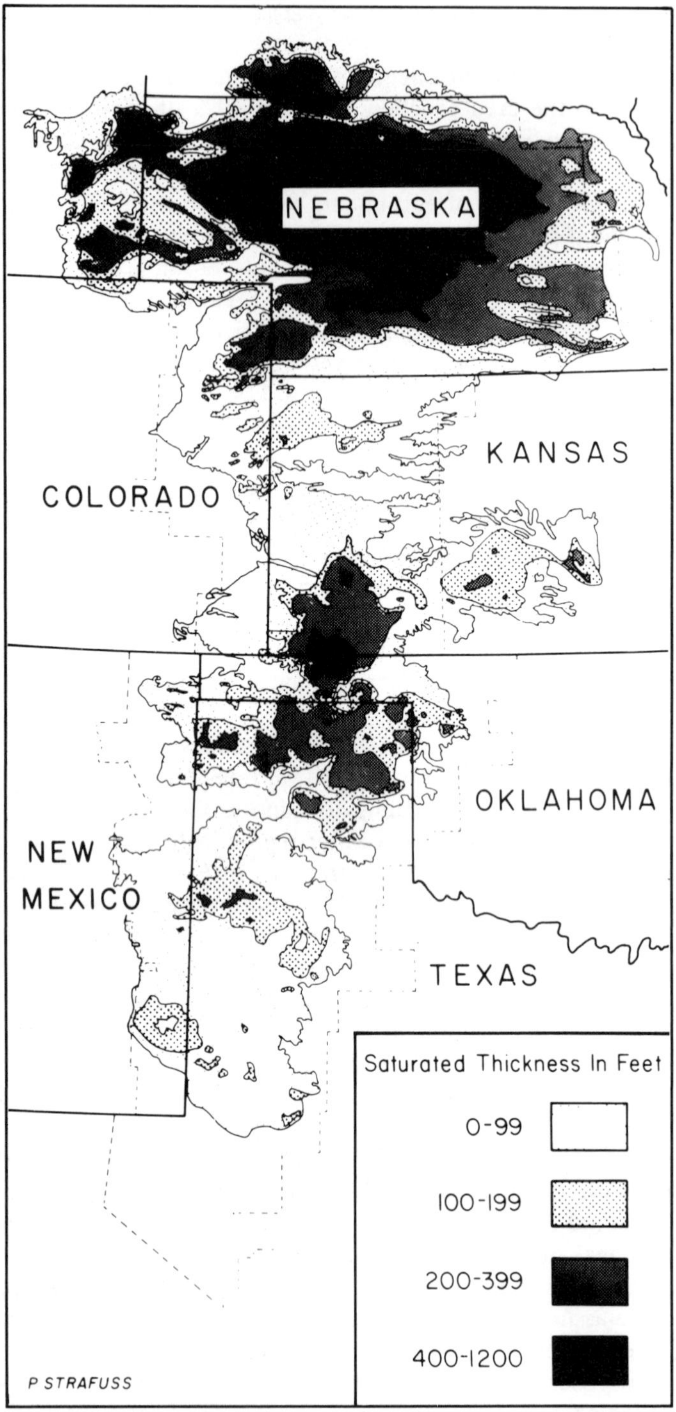

Fig. 9.1. Saturated thickness of the High Plains Aquifer (Kromm and White, 1987). (Reprinted with permission from *Journal of Geography*, 86, 1987.)

millions of hectares of land by the federal government. Despite heightened social and environmental awareness, the people of the Great Plains, even of the most economically deprived counties, are as yet unwilling to believe they have been defeated, and willing to give their land back to the buffalo. Great Plains residents await an alternative solution to the problems of the region.

One of the most graphic examples of the intensification of development at the expense of the rural system is with regard to the use of agricultural chemicals. The United States accounted for 26% of the world pesticide market in 1987, spending US$6.9 billion and using more than 0.9 billion tonnes of active ingredients; 75% of pesticides were used for agriculture (Harger, 1990). It is simply not in the typical farmer's interest to change practices. Federal crop insurance policies and tax law provisions hinder wide-scale adoption of some environmentally and economically sound techniques. At one end of the 'alternative agriculture' spectrum are the organic growers, an estimated 1–2% of the nation's 2.2 million farmers, according to slightly varying legal definitions in 14 states. At the other end of the spectrum are the low-input advocates. Motivated by the need to save money, an estimated 30–40% of American farmers have taken simple steps to reduce their use of chemicals. The information on low-chemical-use strategies is also lacking. According to a poll by Iowa State University, 76% of the state's farmers believe that modern agriculture relies too heavily on chemicals, and 56% would adopt low-input methods if more information were available. But farmers lack information because many researchers at land-grant universities focus on projects that will be funded by agrochemical companies (McDermott, 1990).

Commodity programmes of the past have encouraged chemical intensive monocultures. Farmers' eligibility for payments under commodity support programmes was proportional to their established area base (the amount of land planted to specific crops in preceding years). Thus farmers were penalized for shifting crops out of supported crops (Faeth *et al.*, 1991). The 1990 Farm Bill has attempted to address this issue.

Future Considerations Influencing Restructuring

The 1990s could be the decade of environmental conscientiousness, although commodity organizations continue to fight recognition of the impact of environmental externalities on agricultural systems. The need for sustainable agricultural systems could lead to a range of responses placing more value on the environment; strengthening the conservation programme, for example, seems likely. Although excess capacity persists in US agriculture, it is not the serious problem it was a few years ago;

conservation reserve lands are likely to expand, although the rate of reservation will probably slow.

As the US moves towards a greater emphasis on sustainability, there will need to be greater efforts at compliance with conservation plans. In Kansas, for example, the state will develop a 'State Water Use Management Program' which will use 'incentives' (such as the linkage of certain state grant awards to the implementation of conservation plans) to encourage water users to adopt and implement good water-management practices (Kansas Water Office, 1990). Similar major replanning in some farm areas may also be required to enforce reductions in erosion.

The farm credit system must also begin to value environmental externalities. As part of this effort, the farm credit system and lenders participating in federally chartered secondary markets must require borrowers to comply with sodbuster, swampbuster and conservation compliance.

Overall, an economic environment must be developed that will encourage sustainable farming. Schemes vary from a complete overhaul of the US Department of Agriculture commodity programmes to flexible programmes that will allow producers to shift cropping patterns without a loss of their base areas. But if environmental issues are to have a higher priority and a greater recognition of their related externalities, and if greater use of low-input production methods is to be achieved, some form of government intervention is necessary (Foulke, 1989). This includes greater participation by state and local governments in addressing such issues as: the population resurgence in some rural communities; wide-scale concerns about the future of the family farm; and growing competition between industry, tourists and environmentalists for control over rural resources.

As we look to the future, the need for a major restructuring of rural systems will no longer be a radical argument, as the National Academy of Science predicted, but conventional wisdom, with significant benefits for farmers, the economy and the environment. A new sense of stewardship comes none too soon: as Will Rogers once quipped – 'Land, they make so little of it nowadays' (Carey, 1986).

References

Carey, J. (1986) 50 years of land use – the changing face of America. *National Wildlife* April/May, 18–26.

Clark, E.H. and Haverkamp, J. (1985) *The Off-farm Costs of Soil Erosion*. The Conservation Foundation, Washington DC.

Cohen, W., Hug, A., Taddese A. and Cook, K. (1991) FACTA 1990: conservation and environmental highlights. *Journal of Soil and Water Conservation* 46, 20–2.

Cook, K.A. (1989) The environmental era of US policy. *Journal of Soil and Water Conservation* 44, 362–6.

Cook, K.A. and Hinkle, M.K. (1989) *Statement before the Subcommittee on Agricultural Production and Stabilization of Prices*. Senate Committee on Agriculture Nutrition and Forestry, Washington DC.

Daniels, T., Lapping, M. and Keller, J. (1989) Rural planning in the United States: fragmentation, conflict and slow progress. In: Cloke, P.J. (ed.), *Rural Land-Use Planning in Developed Nations*, Unwin-Hyman, London, pp. 152–77.

Dawson, P. (1988) Deborah and Frank Popper, the fate of the Plains. *High Country News* 20, 16–20.

Drabenstott, M. and Barkema A.D. (1990) US agriculture charts a new course for the 1990s. *Economic Review* January/February, 32–49.

Ervin, D.E. (1988) Set Aside Programmes: using US experience to evaluate UK proposals. *Journal of Rural Studies* 4, 181–91.

Faeth, P., Repetto R., Kroll, K., Dai, Q. and Helmers. G. (1991) *Paying the Farm Bills: US Agricultural Policy and the Transition to Sustainable Agriculture*. World Resources Institute, Washington DC.

Foulke, J. (1989) Low-input farming faces profitability issue. *Farmline* 2,13.

Gardner, P. and Wood, W.W. (1984) Agricultural land use policy: implications for state and local government. *California Agriculture* July–August, 6–8.

Harger, C. (1990) The greening of America's heartland. *Conservation* 8, 6–11.

Kansas Water Office (1990) *Kansas Water Plan Executive Summary*. Kansas Water Office, Topeka.

Kirschenmann, F. (1991) Fundamental fallacies of building agriculture sustainability. *Journal of Soil and Water Conservation* 46, 165–8.

Kromm, D.E. and White, S.E. (1987) Interstate groundwater management preference differences: the Ogallala Region. *Journal of Geography* 86, 5–11.

Kromm, D.E. and White, S.E. (1990) *Conserving Water in the High Plains*. Kansas State University, Manhattan.

McDermott, J. (1990) Some heartland farmers just say no to chemicals. *Smithsonian* 21, 114–27.

Nellis, M.D. (1987) Land use adjustments to aquifer depletion in Western Kansas. In: Cocklin, C., Smith, B. and Johnston, T. (eds), *Demands Upon Rural Lands: Planning for Resources Use*. Westview Press, Boulder, CO, pp. 71–84.

Nielsen, E.G. and Lee, L.K. (1987) *The Magnitude and Costs of Groundwater Contamination from Agricultural Chemicals: A National Perspective*. AERS 576. Economic Research Services, USDA, Washington DC.

Pins, K. (1986) Poll: Iowans want limits on ag-chemicals. *Des Moines Sunday Register* 16 Nov.

Steiner, F. (1987) Viewpoint. *Planning* 53, 46.

Young, C.E. and Osborn C.T. (1990) Costs and benefits of the Conservation Reserve Program. *Journal of Soil and Water Conservation* 45, 370–3.

10

Natural Heritage and Agricultural Production in Canada

Stewart Hilts

The relationship between natural heritage features and agricultural production in the rural landscape is changing rapidly in all regions of Canada. At the most general level, this change involves increasing recognition of environmental values and some de-intensification of agriculture. Co-operative approaches to achieving this shift in perspective imply the development of a new partnership between agricultural and environmental interest groups, but the defensiveness of the farming community could undermine this trend.

Discussions on the environmental impact of agricultural activities in Canada involve a number of particular issues, including soil erosion and water quality degradation, the over-use of agricultural chemicals and fertilizers, and the loss of wildlife habitat and other natural heritage values. In this chapter the relationship of agriculture and natural heritage is examined. First, the extent to which past agricultural programmes have ignored environmental concerns is briefly noted. Then three reasons for a shifting balance in this relationship are described: (i) widespread increasing concern over environmental issues in general, and in agriculture specifically; (ii) the income crisis in farming; and (iii) the development of new 'private stewardship' approaches to rural land policy. In the latter case three specific programmes are described as illustrations.

The result of these changes could be a fundamental shift in the focus of farm subsidy programmes from production objectives to a more balanced mix of production and environmental objectives. In other words, Canadian policy is beginning to move from seeing farmers entirely as producers of food, to seeing farmers as stewards of rural land, one product of which is food. Such a trend seems to have occurred in Europe (see Chapter 15) and the United States (see Chapter 9) in advance of Canada; experience in these cases is also influencing the direction of agricultural policy in Canada.

Conventional Agricultural Policy in Canada

Over the past 20–30 years and longer, agriculture has been promoted as the dominant user of the settled rural landscape in Canada, to the exclusion of natural heritage and other environmental values. Small woodlots and wetlands, water resources, and many wildlife populations have declined or been lost as agricultural production has intensified. Relevant agricultural policies have not merely failed to protect heritage values, they have actively encouraged their destruction.

Agricultural policy to date has largely been based on production and income goals, with a tendency to encourage large-scale monoculture of crops, and an inherent short-term economic and political perspective (see Chapter 15). These policies have, in Dumanski's words, given farmers '. . . incorrect signals relative to land stewardship and the environment' (1988, p.77). In the case of southern Ontario, for example, subsidies have been available for drainage of wetlands and removal of forested areas; in the prairie region of Canada, farmers have been encouraged to fill in and drain the thousands of small wetlands known as potholes.

A recent national review of agricultural policies and wildlife habitat concluded that the environment has largely been ignored in this context, in favour of economic priorities. In fact, agricultural policies in Canada were described as working at '. . . cross-purposes with objectives for the environment and wildlife habitat' (Girt, 1990, p.vi).

Some authors would make this case even more strongly. MacNeill (1989) has generalized beyond Canada, and argued that western governments have established an agricultural subsidy structure that encourages the destruction of resources that should be maintained for effective soil and water conservation. Forest and woodlot clearance, the use of marginal land for production, the over-use of pesticides and fertilizers, and the wasteful use of water all result from typical agricultural subsidies. Our agricultural subsidy structure actually leads to the deterioration of our soil and water resources.

Pointing out that '. . . destruction of the planet's basic environmental capital . . . [is] driven not only by private decisions but also by public policies, often backed by huge budgets' (MacNeill, 1989, p.52), MacNeill proceeds to condemn agricultural policies and budgets for their focus on production, though he does not blame the farmers themselves.

As he describes the situation, farmers work within the context set by policies and available subsidies, even when the signals of government programmes are contradictory. The simple fact is that enormous subsidies for production dwarf any small grants available to encourage conservation.

The dominant message of Canadian agricultural policy has clearly been focused on food production goals, and has largely ignored broader, especially off-farm, environmental effects. Even agricultural conservation programmes, based largely on the principle of voluntarism, have been criticised for failing to lead to the widespread adoption of soil and water conservation practices (Lovejoy and Napier, 1986). But in the face of a dominant production message, is this surprising?

Patterns of Change

Three forces are now combining to encourage a change in this pattern. The first is widespread increasing recognition and concern over environmental issues in general, and the environmental impact of agriculture specifically – in some people's eyes, upsetting the very sustainability of agriculture. A soil and water conservation 'crisis' has been a major contributor to this emerging concern in Canada.

Second is the income crisis of the farming community. Low world crop prices and rising costs squeeze farmers in Canada as elsewhere in the western world.

The third significant influence is the emergence of a number of 'private stewardship' programmes: practical programmes designed to encourage good land stewardship practices by individual landowners through the use of moral and economic incentives. Such programmes are a small but important change in the rules of the game for all rural landowners, and are being received very positively in a variety of Canadian pilot projects. Specific examples from the Prairie region and from southern Ontario are described below.

The sustainability of agriculture has become the centre of a major debate in Canada, as politically the nation has taken a turn toward the environment. Canada's 'Green Plan', put forward by the mainstream Conservative party, although limited in the eyes of many environmentalists, shows just how central the issue has become (Government of Canada, 1990). The environmental sensitivity of programmes is now being examined in all sectors, and agriculture is no exception.

Probably the most dominant environmental issue within the farm community itself has been the soil and water conservation 'crisis'. A well-publicized series of hearings by a national Senate Standing Committee, chaired by Senator Sparrow, produced a widely distributed report entitled *Soil at Risk: Canada's Eroding Future* (Senate Standing Committee, 1984). This report, and Senator Sparrow's many personal appearances to spread its message, has galvanized agricultural ministries more than anything else, with the result that new government programmes to encourage soil conservation have appeared across the country.

The debate has become much broader than this though, as various views of 'alternative agriculture', including organic farming, have been put forth. The most popular term now is clearly 'sustainable agriculture', perhaps because even the most intensive conventional farming can make the claim of being sustainable. Recently both a House of Commons Standing Committee and a Science Council Committee have conducted national reviews of 'sustainable agriculture'.

Within this debate, much of the concern revolves around on-farm environmental impact, for example, soil erosion that is severe enough to reduce farmers' own crop yields. Presumably any farmer has a personal incentive to avoid environmental damage on his own land, and much of the educational material accompanying present soil and water conservation programmes emphasizes the on-farm benefits. Increasingly though, off-farm costs, such as downstream watercourse pollution and wildlife habitat loss, are being recognized. This raises the debate over the environmental impacts of agriculture to a different level, since society at large has a stake in these off-farm environmental impacts (Stonehouse and Bohl, 1991).

In sum, this debate over the sustainability of agriculture itself, and increasing recognition of the off-farm environmental impacts of agriculture, is leading both policy-makers, and the farm community itself, to search for new directions that could make agriculture more sensitive to the environment. This will be one major influence on rural restructuring in the coming years.

The second major element of this pattern of change is the income crisis in farming. There is no need to repeat the growing litany of stories about farm bankruptcies, low world grain prices combined with high input costs, and other stresses that combine to make farming a low profit occupation for farmers in Canada (see Chapter 2). Subsidy programmes, both in North America and Europe, play a major role in creating this situation. Canada, as a relatively small player in the world market, is subject to the economic context established in the world at large. At the same time, though, subsidies provide a large proportion of net farm income.

Some farm community leaders have seen the connection quite quickly. On the one hand, farmers want more public financial support; on the other hand the public wants more environmental sensitivity from agriculture. Putting the two together suggests the obvious idea of providing subsidies for a change of environmentally appropriate practices by farmers. If farmers can balance their production objectives with conservation objectives, the result will be greater public acceptance of farm subsidy programmes.

Dumanski (1988) described the idea succinctly 4 years ago. Noting that society will demand greater environmental sensitivity from agriculture in the future, these authors suggested that sustainable stewardship of our land will increasingly be seen as the responsibility of the farm community. The appropriate goal for agriculture then would be to generate tangible public support to share the costs of conservation programmes. Simply put, Dumanski (1988, p.77) suggested that '. . . programmes with broad conservation objectives can expect greater political and financial support from a wider segment of the public'.

One current example in Ontario is the proposal to lease or purchase conservation easements from farmers in the Niagara Fruit Belt, under intense pressure from urban development, providing a major financial incentive in return for permanent protection of the agricultural land base (Task Force on Agricultural Finance and Farmland Issues in the Niagara Tender Fruit Industry, 1991). A second example is a suggestion to tie the provincial farm property tax rebate system, whose cost has been skyrocketing, to requirements for a land stewardship plan for a farm (van Donkersgood, 1989a).

Similar arguments have been raised in favour of 'decoupling', that is the separation of farm subsidies from production of specific commodities. Van Donkersgood (1989b) suggests that the goal of government programmes should be a balance of production and stewardship, that measuring success by production alone is not enough. Rural landowners should also have the responsibility of passing the land '. . . on to the next generation in as good a condition as it was received from the previous generation' (1989b, p.15).

This view of 'decoupling' must be distinguished from one that argues simply for a return to more free market conditions; but the point is that as farmers search for alternatives in the face of a farm income crisis, any reason to justify greater public support is welcome, and subsidies for environmental stewardship may be as welcome as for any other reason.

A third significant development in recent years is the emergence of a variety of 'private stewardship' programmes in Canada (Hilts, 1990). In one sense, these programmes are simply renewed versions of past farm extension programmes where direct one-to-one contact is made with a landowner, to encourage their adoption of up-to-date farm practices. However, these stewardship programmes have some differences. In some cases, direct financial incentives are now available to encourage participants, in the case of soil conservation, for quite substantial amounts. And the new stewardship programmes have as their central goal soil and water conservation, or the protection of wildlife habitat and natural heritage values, rather than production goals.

The province of Ontario provides two examples. On the agricultural side, the soil and water conservation programmes known as Land Stewardship I and II are providing direct grants to farmers for a range of specific conservation options, including the adoption of conservation tillage practices. The federally funded National Soil Conservation Programme now offers complementary incentives for such practices as the reforestation of highly erodible land and streambank margins. Under these programmes, grants can presently run as high as C$10 000 – certainly a substantial incentive for any farmer. An interesting aspect of these programmes is that they are both administered through a cooperative agreement with the Ontario Soil and Crop Improvement Association, and at the local level farmers have a large measure of control over the granting decisions.

These programmes are proving extremely popular among Ontario farmers, although they are not without their problems, most of which revolve around equity questions in the administration of the grants. In these cases, farmers are readily accepting public financial support for practising environmental stewardship on their farms. A total of approximately C$88 million will be spent in these programmes over 6 years ending in 1993 (Moull, 1990).

A second private stewardship programme in Ontario, with which the author has been directly involved for 8 years, has the goal of protecting natural heritage values in the rural landscape. In the National Heritage Stewardship Programme landowners are contacted and encouraged to make a personal or ethical commitment to conservation, through a verbal, voluntary or 'handshake' stewardship agreement. Reaction from landowners has been very positive, although their long-term commitment is still in some question. Along with this, a small financial incentive is offered, in the form of a property tax rebate, under the Conservation Lands Tax Reduction Programme. This is also proving successful, although there is often considerable reluctance to make a written commitment to conserving woodlots or wetlands in return for such a rebate, because of future possible development potential (Van Patter *et al.*, 1990).

This particular programme has only limited financial incentive to offer farmers, who are usually not eligible for the conservation tax rebate programmes because they already receive a similar farm tax rebate. But the idea of providing larger incentives is now under discussion, and landowner interest in such schemes seems high.

A third example comes from the Prairie provinces: Manitoba, Saskatchewan and Alberta. Here the focus of the relevant private stewardship programmes has been on preservation of 'prairie potholes' – the thousands of small wetlands that dot parts of the prairie landscape and provide nesting habitat for much of the duck population of North America. Duck populations have been in rapid decline over the past decade, leading the hunting fraternity to support increasingly active efforts to conserve these wetlands. In several pilot projects jointly funded by provincial governments and Wildlife Habitat Canada (a national conservation organization), farmers are being offered a range of financial payments in return for leases over potholes and surrounding nesting habitat (Melinchuk, 1987).

As in the Ontario cases, these projects have been quite successful in attracting participation by farmers. They are another example of the direct payment of financial incentives for environmental stewardship by farmers. The prospect of large-scale funding for wetland conservation under the North American Waterfowl Management Plan, a joint US/Canadian programme to raise C$1.5 billion in conservation funds over 15 years, may enable quite widespread payments to farmers, especially in the Prairie provinces.

These stewardship projects are based on cooperative partnerships between conservation agencies and landowners, rather than on a regulatory relationship. They need to be implemented through the action and commitment of local rural communities. In the immediate future a number of policy options, from voluntary moral persuasion through to economic incentives and stronger legal agreements, will be tried and evaluated in different combinations. The net result of this will amount to a shift in agricultural subsidies from being dominantly based on production goals to being at least partly based on environmental goals.

Such a basic shift in policy is the result of the general increasing public interest in environmental issues, the readiness of farmers to consider other income sources, and the development of these stewardship programmes.

Discussion

This trend in policy development in Canada may contribute significantly to overall rural restructuring, but it has been preceded by developments in both the United States and in Europe. The US Farm Bill, with its revolutionary cross-compliance requirements, is a major precedent in North America (Cook, 1989). As governments move toward improved policy integration, various forms of cross-compliance will be examined in an effort to minimize the expenditure of new tax dollars, and achieve desired results more efficiently.

European precedents include several programmes in different countries, but tend in general to reflect a position that rewards the farmer for environmental stewardship as well as production goals (Grossman, 1989; OECD, 1989). Some of these, such as the Environmentally Sensitive Areas programme, are quite specific in purpose (O'Riordan, 1989; Baldock *et al.*, 1990), but they appear to reflect quite a different attitude toward the role of the farmer. And to some extent at least, farmers themselves are involved with promoting this wider perspective, such as in the case of Farming and Wildlife Advisory Groups in Britain (Cox *et al.*, 1985).

Although Canadian attitudes and policies do not yet reflect this broader view of the farmer as landscape steward, as well as food producer, it appears that the trend is in the same direction. Certainly increasing efforts will be made in the near future to ensure that national and provincial agricultural policies do not result in inadvertent destructive impacts on environmental values.

The major constraint to this shifting balance is the defensiveness of the agricultural industry and bureaucracy, and to a lesser extent some rural farm communities. In some cases serious conflict between rural and ex-urban residents makes co-operation difficult to achieve. Especially in regions where the farm income crisis is more serious, and where recreational use of rural land by ex-urbanites is increasing, farmers tend to react negatively to suggestions for improved land conservation. In some cases, such protests have reached a near violent level.

One possible result is that farm communities will not accept the validity of a broader view of the rural landscape, and conflict will surround the continued pattern of change in rural areas. Such defensiveness, in the face of increasing taxpayer willingness to provide subsidies, will weaken the agricultural community in Canada in the long run, and at the same time lead to the continued loss of natural heritage and other environmental values.

Achieving a shift in this relationship through positive private stewardship programmes and innovative policy change – the more cooperative route – will enable the creation and maintenance of a more ecologically sustainable landscape and more socially sustainable rural communities.

References

Baldock, D., Cox, G., Lowe, P. and Winter, M. (1990) Environmentally Sensitive Areas: incrementalism or reform? *Journal of Rural Studies* 6, 143–62.

Cook, K.A. (1989) The environmental era of US agricultural policy. *Journal of Soil and Water Conservation* 44, 362–6.

Cox, G., Lowe, P. and Winter, M. (1985) Land use conflict after the Wildlife and Countryside Act 1981: the role of the Farming and Wildlife Advisory Group. *Journal of Rural Studies* 1, 173–83.

Dumanski, J. (1988) A Canadian perspective on the Food Security Act. *Journal of Soil and Water Conservation* 43, 76–7.

Girt, J. (1990) *Common Ground.* Wildlife Habitat Canada, Ottawa.

Government of Canada (1990) *Canada's Green Plan.* Supply and Services Canada, Ottawa.

Grossman, M.R. (1989) Farmland and the environment: protection of vulnerable agricultural areas in the Netherlands. *Agriculture and Human Values* 6, 101–9.

Hilts, S. (1990) Private stewardship: its beginnings and use across Canada. In: Nelson, J.G. and Woodley, S. (eds), *Heritage Conservation and Sustainable Development.* Heritage Resources Centre, University of Waterloo, Waterloo, Ontario, pp. 191–6.

Lovejoy, S.B. and Napier, T.L. (1986) Conserving soil: sociological insights. *Journal of Soil and Water Conservation* 41, 304–10.

MacNeill, J. (1989) The costs of inaction: what can and should be done and by whom? In: Davidson, D. and Dence, M. (eds), *The Brundtland Challenge and the Cost of Inaction.* Institute for Research in Public Policy, Dalhousie, Halifax, pp. 51–64.

Melinchuk, R. (1987) Saskatchewan's prairie pothole project. In: *Wildlife Conservation on Private Lands*. Wildlife Habitat Canada, Ottawa, pp. 10–20.

Moull, T. (1990) Private stewardship programs in Ontario. In: Nelson, J.G. and Woodley, S. (eds), *Heritage Conservation and Sustainable Development*. Heritage Resources Centre, University of Waterloo, Waterloo, Ontario, pp. 209–16.

OECD (1989) *Agriculture and Environmental Policies: Opportunities for Integration*. Organization for Economic Cooperation and Development, Paris.

O'Riordan, T. (1989) Changing landscapes and land use patterns and the quality of the rural environment. In: *Agricultural and Environmental Policies: Opportunities for Integration*. Organization for Economic Cooperation and Development, Paris, pp. 169–80.

Senate Standing Committee [on Agriculture, Fisheries and Forestry] (1984) *Soil at Risk: Canada's Eroding Future*. Supply and Services Canada, Ottawa.

Stonehouse, P. and Bohl, M. (1991) Land degradation issues in Canadian agriculture. *Canadian Public Policy* (in press).

Task Force on Agricultural Finance and Farmland Issues in the Niagara Tender Fruit Industry (1991) Report to the Honourable Elmer Buchanan, Ontario Minister of Agriculture and Food, 16 May, 1991 (mimeograph).

van Donkersgood, E. (1989a) Decoupling: a family farm and stewardship approach. *Earthkeeping* 5, 12–5.

van Donkersgood, E. (1989b) A case for cross compliance. *Earthkeeping* 5, 28.

Van Patter, M., Geerts, H. and Hilts, S. (1990) Enhancing private land stewardship. *Journal of Natural Areas* 10, 121–8.

IV

The Development of Policy and Programmes

11

The Convergence of Agricultural and Environmental Policies: The Case of Extensification in Eastern England

John Tarrant and Richard Cobb

The last 10 years have seen considerable changes in policy towards both agriculture and the agricultural landscape. These changes, although much heralded, particularly by those concerned with agricultural policy, have been neither rapid nor particularly successful. The changes derive from two different pressures which, while being traditionally opposed, are now moving towards similar policy instruments to meet different objectives. The first is the pressure to reform the Common Agricultural Policy (CAP) of the European Community (EC); the second is the pressure to counteract the damaging environmental effects of intensive agriculture.

The CAP and Environmental Policy in the UK

There are external pressures to change the CAP, from the Cairns Group, from the United States and from the GATT negotiations, but the most significant pressures, and the ones that are forcing the pace, are internal. Policies devised to maximize agricultural production, and thereby to achieve self-sufficiency in temperate products, made economic and political sense in the early years of the Community's history. These policies have to be seen, not in the light of today's problems, but in the context of the need to complete the reconstruction of the European economies after the Second World War.

The resulting support for Community agriculture proved very successful in increasing production and achieving self-sufficiency. However, the policies have proved exceptionally difficult to put into reverse once their objectives had been reached (Fennell, 1985; Bowler, 1986; Franklin, 1988). At the same time the political and economic costs of continuing as before have become unacceptable (Body, 1982; Blunden and Curry, 1985; Commission of the European Communities, 1985a; Cox *et al.*, 1989; Howarth, 1990; Gilg, 1991). In particular the costs of maintaining high prices to producers, and the storage and ultimate disposal of that part of the resulting production which is surplus to European demand at these high prices, have proved unsupportable.

Therefore, from the agricultural policy side we have seen price reductions (in real if not always in absolute terms), price stabilizers where the level of price support falls as aggregate production exceeds fixed quantities, production quotas and finally schemes to encourage diversification out of the products that are in surplus, and even out of agriculture altogether (MAFF, 1987a; 1987b; Marsh, 1987; Blunden and Curry, 1988; Ilbery and Stiell, 1991; Shucksmith and Winter, 1991).

While these changes have been accelerating, there has been rising concern about the environmental effects of the styles of intensive agriculture that have been encouraged by guaranteed high support prices (see Chapter 10 for a Canadian comparison). Increased intensity of production, characterized by reduced use of labour and increased investment in machinery, and increased use of chemicals, have produced environmental change involving a reduction in habitat diversity through the removal of hedgerows and woods, the drainage and ploughing of meadows and marshes, and damage through the pollution of surface and ground waters (Bowers and Cheshire, 1983; Adams, 1986; Lowe *et al.*, 1986; McInerney, 1986). These changes are not new. Concern about field enlargement and hedgerow removal was the subject of much academic work in the 1960s (Lowe *et al.*, 1986), and investment in machinery in United Kingdom agriculture has been declining in real terms from a peak reached some 15 years ago (Burrell *et al.*, 1990). It is perhaps ironic that the pace of change in the enactment of environmental policies for the protection of the countryside has accelerated just as the rate of change wrought by agriculture has probably slowed.

Policies for the environment

Policies to reduce environmental change and damage were, at first, site specific. Farm management agreements, introduced by the 1968 Countryside Act, allowing payments for management work deemed necessary on Sites of Special Scientific Interest (Blunden and Curry, 1985), were expanded under the Wildlife and Countryside Act of 1981. Farmers could now be compensated for the loss of their anticipated future profits foregone where 'improvement' of land was precluded under a management agreement (Benson and Willis, 1988; O'Riordan, 1990; Robinson, 1990; Gilg, 1991). Considerable problems over the implementation of the 1981 Act led to the introduction of Environmentally Sensitive Areas (ESAs) under EC regulation 797/85 (Commission of the European Communities, 1985b). The 1986 Agriculture Act allowed the Ministry of Agriculture, Fisheries and Food (MAFF) to designate areas where landscape and ecology are critically dependent on the maintenance, or adoption, of particular, usually extensive, farming methods (Potter, 1988; Baldock *et al.*, 1990). The idea of the ESAs has been taken up and expanded recently by the Countryside Commission and its Countryside Stewardship Scheme. Ten-year agreements will be offered in carefully targeted areas with the most ecological and landscape value, particularly chalk and limestone grassland, lowland heaths, river valleys, coastal areas and uplands (Countryside Commission, 1991). Although the effects of such schemes are locally valuable, they have little impact on the general agricultural landscape, and almost no effect on the problems of structural surplus production in European agriculture.

Setaside

The economic and political requirement to reduce the costs of overproduction, and to reduce and reverse environmental damage, came a little closer together with the design of a Setaside Scheme. Under setaside, land producing products for which there is a common market organization within the EC can be taken out of production with the payment of compensation to the farmer (MAFF, 1988; Ilbery, 1990). The significance of this scheme, introduced into the UK in the 1988/89 crop year, was that withdrawal of land could take place wherever it was. The results were not to be concentrated in areas of high landscape and conservation value. Whereas the ESAs have the potential to be of much more value as conservation instruments, setaside can have much more effect on agricultural production. Land withdrawn from production for 5 years, and then reploughed, is of little conservation significance (Bowers, 1987; Burnham *et al.*, 1988). Setaside, of little conservation value, has moved further from conservation objectives following the EC budget crisis of 1991. Land can now be setaside for 1 year only. Farmers have to agree to remove only 15% of their land from production instead of the 20% under the full scheme. Although lower compensation payments are on offer, adoption of the scheme allows farmers to avoid a co-responsibility levy on all the cereals they grow, which makes the scheme more attractive on the large farms. A top-up, or premium scheme, has encouraged farmers to use the setaside land more effectively for the benefit of wildlife, landscape and the local community, and is presently managed by the Countryside Commission in the Eastern Counties, but will be applied nationally and managed by MAFF from 1992. This scheme is applicable only where farmers have agreed to fallow the land for 5 years (Countryside Commission, 1989). The conservation aspect of setaside may be strengthened as, in those areas where ecological value is significant, it is proposed that setaside will be for periods of up to 20 years with a special premium added for environmental protection (Commission of the European Communities, 1990, 1991).

Extensification

Arguably, the objectives of agricultural and environmental policy came closer together with the introduction of the concept of 'extensification'. On the initiative of the EC (Regulation 1094/88) a scheme was devised to lead to a reduction of 20% in agricultural production for at least 5 years through the encouragement of less-intensive styles of farming (Commission of the European Communities, 1988). The reduction is not to be achieved through taking land out of production but payments can be made, either for a drop in output however else achieved, or for the adoption of production methods that under normal circumstances would be expected to reduce production by 20%. It can be used anywhere, in areas of high conservation value and/or in areas of high agricultural production. In principle, therefore, it is a policy that addresses the problems of surplus agricultural production *and* general landscape improvement throughout the Community. Farmers can gain through reduced costs and by incentive payments to compensate for lost production. The Community and its taxpayers should gain as surplus production is reduced and the environment should gain through a slowing down of change.

An Investigation into Agricultural Extensification

The study reported in this paper concentrates on extensification. A consultation paper on a pilot scheme for extensification in beef and sheep production was issued by MAFF in June 1989 (MAFF, 1989). This paper included a commitment to issue a similar consultation paper on an arable extensification scheme 'in due course'. The research started on the assumption that the scheme would be introduced by 1990 but it appears to have been shelved. Meanwhile the EC has indicated that revisions to the scheme are to be announced to link it specifically with other schemes for landscape protection and to make it more enforceable, probably by making payments only on the achievement of reduced production rather than on the basis of adoption of methods that might reduce it. It is possible that extensification will be adopted within the EC as the first tier of the ESA scheme, with farming in the designated ESAs qualifying for additional payments (Commission of the European Communities, 1990, 1991).

The farm sample and survey

If extensification is to be successful in having a significant effect on both the problems of agricultural surpluses and environmental damage in the UK, it has to be taken up on a significant number of the large farms in the major cereal-producing areas. If marginal cereal producers, either economically and/or spatially marginal, take up the scheme there will be little reduction in surplus agricultural production. If all the cereal producers in the western and northern parts of the UK stopped production completely, overall cereal production would not fall by the required 20%. Similarly, to have a significant environmental effect, extensification needs to be adopted over a wide geographical area, and especially where the greatest damage is, or has been, wrought – where agriculture has become most intensive.

For this reason a study area in East Anglia, part of the cereal heartland of the United Kingdom, was chosen (Fig. 11.1). If extensification is not adopted here one would expect little success in achieving the agricultural objectives of the policy. In addition, however, it is important that it is the farmers who control the use of the bulk of the land area who are persuaded to adopt the policy. If a number of small farmers, scattered throughout

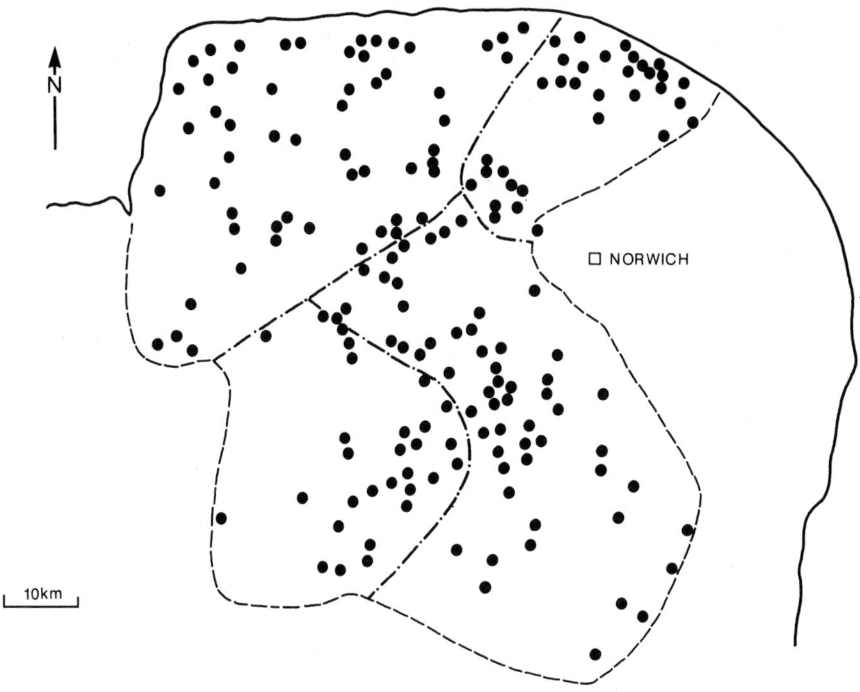

Fig. 11.1. Location of the sampled farms.

the region, adopt more extensive methods, very little will be achieved in terms of production reduction, and there will be only a series of small islands of environmental improvement of very limited conservation or landscape interest. Not only, therefore, has this study concentrated on a major cereal producing area, but it has also examined the reactions of the largest farmers – in what others have called the accumulator class (Munton *et al.*, 1987). The sample of 175 selected farmers in Norfolk and north Suffolk includes over 65% of farmers who control more than 1730 acres (700 ha), but an insignificant number of those controlling less than 124 acres (50 ha). The sample is clearly and deliberately biased towards the farmers with large amounts of land. A high response rate was achieved and the sample includes farmers who together control about 20% of the agricultural land in the study area.

Overall acceptability

The overall reaction to extensification proposals was neither enthusiastic nor realistic. As the payments are designed to compensate for 20% less production resulting from less-intensive styles of farming, a realistic level of compensation would be set at about 20% of the net margin. On most of the cereal land of East Anglia this would not be much more than £50 per acre (£124 per ha). Only 3% of the sample would look favourably on the scheme at this level of compensation, and only 21% responded favourably if the compensation was allowed to rise to £75 per acre (£185 per ha): 30% would not accept the scheme at any price, 17% were undecided, and 32% would be expecting compensation of £100 or more an acre. Even allowing for the difficulties of inquiring about future behaviour among farmers, a position forced on this study as a result of the non-publication of the official government scheme, this result must be regarded as disappointing.

The regional analysis

The study area falls into four subregions on the basis of land capability and historical agricultural practice (Fig. 11.1).

1. The Northeast where the agricultural land is of good quality, predominantly Grade 2.
2. The Northwest where the land is predominantly Grade 3 and where malting barley has become an important crop.
3. The Breckland where the land is very light, the soil being derived from sands overlying chalk. Here farms are large, often with extensive irrigation systems assisting the cultivation of vegetables, especially carrots. These vegetables are rotated with cereals.

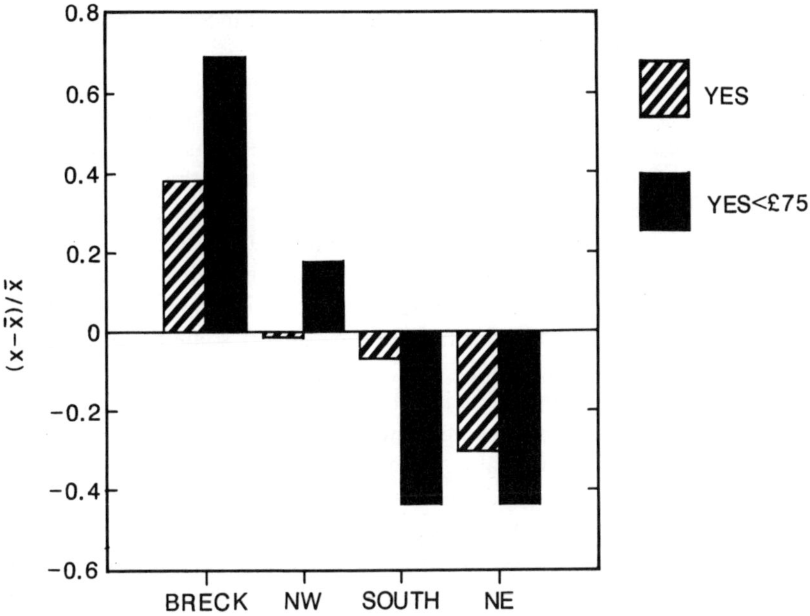

Fig. 11.2. The regional pattern of acceptability of extensification.

4. South Norfolk and north Suffolk where the land is heavy and late, with the soil derived from boulder clay. Here the established Norfolk rotation of wheat and/or barley with sugar beet has recently been supplemented by oil seed rape and field beans.

The regional pattern of the acceptability of extensification is highly significant (Fig. 11.2). It is most acceptable in the Breckland. Here cereals are used as a break crop between production of vegetables. Yields of cereals are already low and payments to reduce the intensity of production (and therefore the costs of production) are attractive. In the Northeast the idea of being paid to reduce production on this high quality land is regarded by many as almost immoral. Between these two extremes lie the South, where the land offers little option but to continue to intensify the existing production of wheat, barley and sugar beet, and the Northwest where there is little enthusiasm for extensification on malting barley land as little nitrogen is used in its production and, therefore, one of the principal ways of reducing inputs is not available. Attitudes to the adoption of extensification at a reasonable price (less than £75 per acre) are similar in spatial pattern to those shown by adoption at any price, but, as would be expected, they are more polarized.

Extensification options and their environmental effects

There are several ways in which extensification might be adopted. The most obvious is to reduce fertilizer and other chemical inputs. A second way is to reduce these inputs and change to a less demanding (and generally lower yielding) crop variety. One way of making this conversion is to turn from winter to spring sown cereals. Third, less land can be cultivated through the use of wider headlands (Deane, 1989) or by introducing a fallowing programme. Aspects of the first and the third methods can be taken together by developing conservation headlands, where field edges are left unsprayed. Finally the farmer can convert some or all of his land to organic styles of production.

The farmers who expressed interest in extensification at £75 or less an acre (£185 per ha) were asked for their preferred method of adopting it. The most popular option (Table 11.1) was to reduce fertilizer use, possibly with some corresponding reduction in the use of other chemicals. Conversion to spring sown cereals would reverse a clear trend in favour of winter cereals. Such a reversal is not popular as spring cereals are viewed as more weather dependent and unreliable and, on the heavier land, early spring preparation of a suitable seed bed is often difficult. The use of conservation headlands was favoured by six farmers, although this technique would not reduce output by the required 20% without being combined with other methods. Fallowed fields and headlands can be included as a part of the setaside scheme. Setaside is unpopular with this group of farmers but some do express a desire for a more flexible setaside scheme that can incorporate fallow and uncropped headlands as a part of an extensification package. It seems probable that payments to reduce the profitability of cultivation in ways that make environmental and agricultural sense, through, for example, rotational fallow, are likely to be more acceptable than payments simply to take productive land out of farming. Organic farming remains a fringe activity; one of the two farmers indicating this as his choice for extensification is already using organic methods on part of the farm.

Little practical work, except on field margins and on small geographical areas, has yet been carried out to establish the environmental benefits of lower input arable farming (Heywood *et al.*, 1988; Doberski, 1989). The lack of a generally applicable incentive scheme for extensification has meant that studies of actual impacts have not been possible, and some examinations of effects have been criticized for poor experimental design and small scale (Sears, 1990). Table 11.1 estimates the likely scale of environmental benefits from various extensification options. The actual benefits will depend on the management of the land (especially the fallow, the headlands and other surrounding uncultivated area).

A further difficulty in assessing these environmental benefits is that

Table 11.1. Extensification options and their environmental effects.[a]

Options	Number of farmers	Wildlife groups and habitats						Watercourse eutrophication	Landscape
		Soil and surface fauna	Arable weeds and invertebrates	Birds			Small mammals		
				Overwinter	Breeding	Predator			
Reduced:									
Fertilizer	18	1	1	1	1	0	0	2	0
Fungicide	6(10)[b]	1	0	0	1	0	1	0	0
Herbicide	2(5)[b]	1	2	1	1	1	1	0	1
Insecticide	0(1)[b]	2	1	2	2	1	2	0	1
Spring sowing	2(3)[b]	1	2	2	1	1	1	1	0
Conservation headlands	6	1	3	1	1	1	1	0	2
Uncropped headlands	4	1	2	2	1	1	1	0	1
Fallowing	6	2	3	3	1	2	2	1	1
Organic farming	2	2	1	2	1	1	1	2	1

[a] Effects:
0 = no effect
1 and 2 = a slowing of environmental damage
3 = some reversal of damage

[b] Figures in brackets include those who might choose this option

Sources: Deane (1989); Heywood (1988); Stanley and Hardy (1984); Sears (1990); Farm survey.

many farmers are already moving in some of these directions as a result of the continued cost-price squeeze. This is especially true for pesticide use where most farmers in the survey are using ‘supervised’ or ‘integrated’ use strategies (Stanley and Hardy, 1984). As with changing farm practices in the ESAs, it is often hard to differentiate changes that farmers would have introduced without further incentive payments from those changes that would be unequivocally a result of introduction of government schemes. Taking the low acceptability of extensification alongside its generally small environmental benefits, it will have little impact on the environmental problems presented by agriculture in Eastern England.

Conclusions

There are three levels of environmental policy applied to the agricultural landscape. At the highest level there are policies designed to reverse the changes wrought by agriculture. These are site-specific and relate to land of the highest conservation and/or amenity value. Within some ESAs farmers are paid to encourage the reversion of farmland to traditional heathland, chalk down grassland, meadow etc. Both the setaside premium and the Countryside Stewardship schemes similarly encourage the restoration of pre-existing habitats. Management agreements were created to prevent change. Payments made to farmers compensate for an agreement not to change the styles of production in ways that would give a higher rate of return but would change the landscape, for example through the drainage and ploughing of wetlands. At the lowest level we have the policy designed merely to slow down, or perhaps halt, the changes through extensification.

The findings of this study can be summarized as follows:

1. Extensification is unlikely to be attractive to farmers who control the majority of the land of East Anglia, at least at realistic levels of compensation.

2. Even where it is adopted, the environmental effects will be limited. A field of wheat will remain as a field of wheat, albeit grown less intensively (Countryside Commission, 1988; North, 1990).

3. In spite, therefore, of a series of attempts including extensification, it seems that an integrated environmental and agricultural policy, covering a significant area of the country, remains elusive.

4. There remains considerable policy confusion between ESAs, Countryside Stewardship Schemes, setaside and extensification. For example, rotational fallow, one of the most promising ways to adopt a more extensive style of production, can be included under setaside, but the proposed levels of compensation for doing essentially the same things are different between the ESAs and the Countryside Stewardship Scheme and between setaside and extensification.

5. Even with a deliberately biased sample drawn only from the 'accumulator' class of farmers, there is significant regional variation in attitudes to extensification. This suggests that the widely adopted classification of farmers into 'accumulator', 'survivor' and 'marginalized' farms (Munton *et al.*, 1987) pays too little attention to the nature of the agricultural land and to the opportunities and restrictions it provides for variety in farm enterprise. Farmers on the best quality land will be the most reluctant to change, whereas those on poorer land may adopt such policies as extensification, not because they are encouraged to adopt changed styles of farming but because they see opportunities to be paid for making changes they were making anyway. Where the land provides little opportunity for diversity, farmers still seem to have little choice but to press on with present, and even more intensive, styles of production.

Acknowledgements

This study was undertaken under a grant from the United Kingdom Economic and Social Research Council within the Council's Joint Agriculture and Environment Programme (JAEP). The authors gratefully acknowledge this financial assistance and the advice of the JAEP steering committees.

References

Adams, W.M. (1986) *Nature's Place*. Allen and Unwin, London.

Baldock, D., Cox, G., Lowe, P. and Winter, M. (1990) Environmentally Sensitive Areas; incrementalism or reform? *Journal of Rural Studies* 6, 143–62.

Benson, J.F. and Willis, K.G. (1988) Conservation costs, agricultural intensification and the Wildlife and Countryside Act 1981: a case study and simulation on Skipworth Common, North Yorkshire. *Biological Conservation* 44, 157–78.

Blunden, J. and Curry, N. (1985) *The Changing Countryside*. Croom Helm, London.

Blunden, J. and Curry, N. (1988) *A Future For Our Countryside*. Blackwell, Oxford.

Body, R. (1982) *The Triumph and the Shame*. Temple Smith, London.

Bowers, J. (1987) Setaside and other stories. In: Baldock, D. and Conder, D. (eds), *Removing Land from Agriculture*. Council for the Protection of Rural England, Institute for European Environmental Policy, London, pp. 5–18.

Bowers, J.K. and Cheshire, P. (1983) *Agriculture, the Countryside and Land Use*. Methuen, London.

Bowler, I.R. (1986) Government agricultural policies. In: Pacione, M. (ed.) *Progress in Agricultural Geography*. Croom Helm, London, pp. 124–48.

Burnham, C.P., Green, B.H. and Potter, C.A. with Edwards, A., Gasson, R. and Shinn, A. (1988) *Setaside as an Environmental and Agricultural Policy Instrument: A Summary Report*. Wye College, Ashford.

Burrell, A., Hill, B. and Medland, J. (1990) *Agrifacts: A Handbook of the UK and EEC Agricultural and Food Statistics*. Harvester Wheatsheaf, Hemel Hempstead.

Commission of the European Communities (1985a) *Perspectives for the Common Agricultural Policy*. COM(85)333, Commission of the European Communities, Brussels.

Commission of the European Communities (1985b) *Council Regulation 797/85 on Improving the Efficiency of Agricultural Structures*. Commission of the European Communities, Brussels.

Commission of the European Communities (1988) *Council Regulation 1094/88 Amending Regulation 797/85 and 1760/87 as Regards the Setaside of Arable Land, and the Extensification and Conversion of Production*. Commission of the European Communities, Brussels.

Commission of the European Communities (1990) *Proposals for a Council Regulation on the Introduction and Maintenance of Agricultural Production Methods Compatible with the Requirements of the Protection of the Environment and the Maintenance of the Countryside*. COM(90)366, Final, Commission of the European Communities, Brussels.

Commission of the European Communities (1991) *The Development and Future of the Common Agricultural Policy – Proposals of the Commission of the European Communities*. COM(91)258, Commission of the European Communities, Brussels.

Countryside Commission (1988) Memorandum of evidence submitted to the House of Lords Select Committee of the European Communities, 10th report. *Setaside of Agricultural Land*. Countryside Commission, Cheltenham.

Countryside Commission (1989) *Countryside Premium for Setaside Land*. Countryside Commission, Cheltenham.

Countryside Commission (1991) *Countryside Stewardship*. Countryside Commission, Cheltenham.

Cox, G., Lowe, P. and Winter, M. (1989) The farm crisis in Britain. In: Goodman, D. and Redclift, M. (eds), *International Farm Crisis*. Macmillan, Basingstoke, pp. 113–34.

Deane, R.J.L. (1989) *Expanded Field Margins. Their Costs to the Farmer and Benefits to Wildlife*. A report to the Nature Conservancy Council, Kemerton Court, Gloucestershire.

Doberski, J. (1989) The Boxworth experiment and the pesticide tightrope. *Ecos* 10, 18–21.

Fennell, R. (1985) A reconsideration of the objectives of the Common Agricultural Policy. *Journal of Common Market Studies* 24, 257–75.

Franklin, M. (1988) *Rich Mans Farming – The Crisis in Agriculture*. Royal Institute for International Affairs, London.

Gilg, A.W. (1991) Planning for agriculture: The growing case for a conservation component. *Geoforum* 22, 75–9.

Heywood, A. (1988) *Cereal Extensification in Lowland England – an Assessment of the Benefits for Wildlife*. Centre for Rural Studies, Royal Agricultural College, Cirencester.

Howard, R.W. (1990) *Farming for Farmers?* Institute for Economic Affairs, London.

Ilbery, B. (1990) Adoption of the arable setaside scheme in England. *Geography* 75, 69–73.

Ilbery, B. and Stiell, B. (1991) Uptake of the farm diversification grant scheme in England. *Geography* 76, 259–63.

Lowe, P., Cox, G., MacEwen, M., O'Riordan, T. and Winter, M. (1986) *Countryside Conflicts*. Gower, Aldershot.

MAFF (1987a) *Farming UK*. Ministry of Agriculture, Fisheries and Food, London.

MAFF (1987b) *Farm Diversification Grant Scheme*. Ministry of Agriculture, Fisheries and Food, London.

MAFF (1988) *Setaside*. Ministry of Agriculture, Fisheries and Food, London.

MAFF (1989) *Pilot Extensification Schemes for Beef and Sheep*. Ministry of Agriculture, Fisheries and Food, London.

Marsh, J.S. (1987) Alternative policies for agriculture in Europe. *European Review of Agricultural Economics* 14, 11–21.

McInerney, J. (1986) Agricultural policy at the crossroads. In: Gilg, A.W. (ed.), *Countryside Planning Yearbook* (7), Geobooks, Norwich, pp. 44–75.

Munton, R.J.C,, Eldon, J. and Marsden, T. (1987) Farmers' responses to an uncertain future. In: Baldock, D. and Conder, D. (eds), *Removing Land from Agriculture*. Council for the Protection of Rural England, Institute for European Environmental Policy, London, pp. 19–30.

North, J. (1990) Oral evidence to the House of Lords' Select Committee on European Communities, 1989/90 session. *24th Report on the Future of Rural Society*, HMSO, London.

O'Riordan, T. (1990) The ever changing politics of nature conservation in the United Kingdom. *Environment and Planning (A)* 22, 143–5.

Potter, C. (1988) Environmentally Sensitive Areas in England and Wales: an experiment in countryside management. *Land Use Policy* 5, 301–13.

Robinson, G.M. (1990) *Conflict and Change in the Countryside*. Belhaven Press, London.

Sears, J. (1990) *A Research Review of Current Information on the Effects on Birds of Organic and Low Input Farming*. Royal Society for the Protection of Birds, Sandy.

Shucksmith, M. and Winter, M. (1991) The politics of pluriactivity in Britain. *Journal of Rural Studies* 6, 429–35.

Stanley, P.I. and Hardy, A.R. (1984) The environmental implications of current pesticide usage on cereals. In: *Agriculture and the Environment*, Proceedings of the Monks Wood Experimental Station, No. 13, Institute of Terrestrial Ecology, Cambridge, pp. 66–72.

12

Agricultural Policies for Urban Fringes in the United States

Timothy Rickard

Public policy to preserve farmland and enhance farm viability on the urban fringe is still evolving in the United States. Despite several decades of research and policy formulation, considerable refinement of theories and programmes is necessary for their effective integration. A number of issues remain problematic: the value of various techniques in preserving farmland is still debated; the diversity of programmes at state and county level across the country makes inventorying and critical analysis difficult; the need for the preservation of agricultural land on the urban fringe is still in doubt, as is its future place among policies to manage growth comprehensively; the particular character of farming on the urban fringe is not sufficiently well understood; and policies to encourage farmers to take advantage of the opportunities for marketing in metropolitan areas are still being developed. This chapter assesses the past decade of scholarship in the US on agricultural policies for the urban fringe and suggests further research issues for the next decade.

The Preservation of Prime Farmland

The debate on the preservation of prime farmland addresses five issues: (i) Why should farmland be preserved nationally, in a state or locally? (ii) What policies and techniques best meet farmland preservation goals? (iii) What policies are currently adopted in states, counties and municipalities in the United States and how do these different government levels work together? (iv) What is the effect of federal policy and what should be the federal role? (v) Is there a second quiet revolution toward comprehensive land management that is diminishing the importance of farmland preservation?

The rationale for the preservation of prime farmland

In 1976, Hart concluded that little more than 4% of the nation's land would be urban by the year 2000 and that urban encroachment would not remove significant areas of land from agricultural production within the foreseeable future (Hart, 1976). Urban encroachment on rural land was not, therefore, a serious problem in the United States; those concerned about a threat to the nation's food supply were unduly influenced by visibly high rates of conversion in the urban fringe of New York City or Los Angeles.

A decade later, Platt (1985) explained that the National Agricultural Land Study of 1979–81 had supported alarmist views of the threat of urbanization to the nation's food supply, which were then discredited by the 1982 National Resources Inventory. He presented, however, the increasingly popular view that there were other reasons for protecting farmland, including efficient and attractive patterns of urban development and protection of a viable local agricultural economy. He noted that high rates of local farmland loss were often obscured by state or regional statistics.

The substitution of state, county or local rationales for preserving farmland in the urban fringe after the mid-1980s effectively fragmented the issue. If there was no national food crisis, then local communities at the different levels had to find their own justifications for farmland preservation. Furuseth (1985) noted that these were often environmental and open space concerns. Critics of farmland preservation policies in California charged that they were surreptitiously conceived schemes to implement no growth policies, with support from elitist community groups and environmental organizations. Daniels (1990) asserted that protecting open space is now 'the prevalent goal in America'. There are various kinds of open space on the urban fringe, however, and subsuming farmland preservation into growth management policies may diminish its importance (DeGrove, 1991).

To justify the restructuring of agriculture on the urban fringe to preserve farmland and farmers, it is still attractive to maintain that their production of food and fibre is worthwhile in the long term. Nelson (1990) has argued that there is some consensus that the preservation of prime farmland at least guards against future uncertainties. It requires fewer inputs relative to marginal land and will someday obviate the need for bringing marginal land into production. Prime farmland nationally is at a premium, a fifth of it being within 50 miles of the nation's 100 largest urbanized areas (Furuseth and Pierce, 1982; Nelson, 1990). Growth management policies should, therefore, give this land a special value.

Techniques and policies for preserving prime agricultural land

Reviews of the large array of available farmland preservation techniques are numerous (Jackson, 1981; Furuseth and Pierce, 1982; Rose, 1984; Atash, 1987; Lapping *et al.*, 1989; Nelson, 1990), but consensus about the effectiveness of the various techniques is still tenuous. There is agreement, however, as to the dearth of rigorous evaluation, particularly by the very states and local governments implementing policy (Daniels, 1990; Nelson, 1990; Coughlin, 1991). Also, the sheer diversity of programmes, each originating in and adjusting to a different political milieu, complicates assessment.

In an overview searching for order and effectiveness in the variety of programmes, Atash (1987) evaluated the major techniques, concluding that incentive programmes, such as differential taxation and agricultural districting, should be integrated with land-use control mechanisms, such as agricultural zoning, in a comprehensive strategy. Separately, incentive programmes or land-use control mechanisms were not effective.

In advancing a more theoretical and prescriptive view of appropriate policy, Nelson (1990) asserted, however, that integrating techniques was not necessarily more effective than adopting one particular technique. He evaluated the varieties of tax incentives and disincentives, right to farm laws, acquisition of development rights and agricultural zoning, concluding that only an urban growth boundary, combined with zoning of prime farmland for exclusive farm use, was appropriate. Demand for rural non-farm dwellings would be accommodated by a 10–20 acre minimum lot size and non-exclusive agricultural zones in areas of non-prime soils. This is a fairly rigorous model, clearly dividing the city from its rural environs (Pease, 1991). Only a buffer of hobby farms would be allowed as, in effect, the urban fringe.

Nelson (1990) also trenchantly dismissed the other techniques widely adopted in the United States as 'subsidizing speculative behaviour, extending the impermanence syndrome across a greater area and generating benefits to the wrong people in the wrong place at the wrong time'. There is some truth to this: tax incentive programmes such as differential assessment (or use-value assessment) do not afford permanent protection and do facilitate speculation; transfer of development rights (TDR) programmes are complicated; purchase of development rights is expensive; agricultural districting encourages speculation and extends the impermanence syndrome, and right to farm laws do not prevent land conversion. On the other hand, farmers on the urban fringe would sell their land sooner if it were taxed for its market value or if they had no protection from neighbours' nuisance complaints. TDR is successful in Montgomery County, Maryland; and agricultural districting works well in areas beyond the urban fringe (Lapping *et al.*, 1989).

Furthermore, evidence in the Northeast shows that over a period of years, a critical mass of preserved farms and farmland can be assembled by the purchase of development rights (Daniels, 1991). Most critiques and theories of preservation techniques and policy in the United States are empirically based. The combination of techniques Nelson advocated derived explicitly from his research on implementation of the 1983 Oregon Land Use Act – perhaps the nation's most celebrated farmland preservation programme (Furuseth, 1981; Gustafson *et al.*, 1982; Daniels and Nelson, 1986; Nelson, 1987, 1988, 1990). Daniels (1990) advocated a different set of elements in a successful preservation programme based on his experience. They are:

> (1) property tax breaks for commercial farmers grossing over US$250 000 a year; (2) agricultural zoning which allows only one three-quarter acre lot per 25 acres; (3) an option to sell development rights to the county or state; (4) right-to-farm laws to protect farmers against nuisance suits for standard farming practices; (5) urban growth boundaries to curb urban sprawl; (6) rural residential zones on lower quality soils and in locations that will not interfere with commercial farming; and (7) protection for farmers against eminent domain.

Daniels wants to implement this programme, which Nelson says only adds techniques to his model, in Lancaster County, Pennsylvania, where he is Director of the Agricultural Preserve Board (Daniels, 1990; Nelson, 1990). The county is on the outer fringe of Philadelphia's urban expansion. Currently five of the seven policies are in effect. Rural residential zones are not being implemented because of nitrate pollution in the ground water, while property taxes are so low that farmers do not even participate in the state preferential taxation programme. Timing appears critical in the adoption of both Nelson's and Daniels' policies. The urban growth boundary must be put in place beyond the urban fringe where there are no growth pressures. Likewise, an exclusive agricultural zone cannot be instituted where there is active land conversion, because farmers will resist the immediate decrease in the market value of their land. The policies advocated by Nelson and Daniels, while ideal for particular localities with far-sighted policy-makers, require considerable adaptation by communities already within the urban fringe.

State and local adoption of farmland preservation policies

Surveys of the adoption of the various farmland preservation policies reveal both spatial diversity and lack of congruence among the state, county and local levels. States differ greatly in the policies they adopt.

Their enabling legislation usually relies on county or local government cooperation for implementation (Lapping *et al.*, 1989).

Farmland preservation at the state level is centred in the Northeast and on the west coast, but includes relatively weak efforts in the South and in the Midwest except for Wisconsin and Minnesota (Daniels, 1990). By now, all states have enacted tax incentive laws and 47 have enacted right-to-farm laws, but Bushwick and Hiemstra (1987) consider a combination of these indirect entry techniques to farmland preservation as the weakest protection attempts. The region where this combination occurs is broadly the southern tier of states: Alabama, Arizona, Arkansas, Louisiana, Mississippi, Missouri, Montana, New Mexico, Oklahoma, South Carolina, Tennessee and Texas. During the 1980s other states adopted more direct methods.

Agricultural zoning, the most widely practised land-use control method, remains fundamentally local, although it is often requested or mandated by state legislation. Exclusive agricultural zoning is rare; non-exclusive agricultural zones with large minimum lot size and area-based allocation are more acceptable on the urban fringe. In 1980, 104 counties and 166 municipalities in 22 states were recorded as practising agricultural zoning by the National Agricultural Lands Survey. Coughlin (1991) presents evidence from particular counties suggesting that the practice has increased greatly, but there is no recent national inventory.

Use-value assessments are also often tied into state programmes. Farmers in Wisconsin, for example, can participate in the state's programme to apply local agricultural land property taxes as dollar-for-dollar credit against state income tax in exchange for signing a restrictive agreement for a minimum of 10 years. Farmers can participate only if the local government has developed agricultural preservation plans and agricultural zones. Similarly, in California, farmers must register in a designated agricultural preserve before receiving agricultural use–value assessment for property taxes (Bushwick and Hiemstra, 1987). State mandated use-value assessments are usually made by the tax assessor of the local municipality which collects the property taxes; this raises issues of fairness when municipalities vary in their enthusiasm for preserving farmland.

Nine states have established purchase of development rights (PDR) programmes: Connecticut, Maine, Maryland, Massachusetts, New Hampshire, New Jersey, Pennsylvania, Rhode Island and Vermont. Various counties elsewhere have also adopted PDR; Suffolk County, New York and King County, Washington, have pioneered much-scrutinized programmes (Daniels, 1990). Transfer of development rights (TDRs) programmes are created solely by local governments. Bushwick and Hiemstra (1987) listed 12 TDRs in northeastern states, only three being active (with Montgomery County, Maryland's programme being by far the most successful).

State and local agricultural districting laws have been adopted in 16 states (Lapping *et al.*, 1989). They are voluntary programmes better suited to rural areas than the urban fringe, in which farmers who organize a district receive a variety of benefits such as property tax relief. New York's programme is the most successful, with one-quarter of the state's land area now being in agricultural districts. One-fifth of this land is in the urban fringe, one-seventh is urban and the remainder rural. The 1971 Act demanded cooperation among state agencies and local governments in setting up agricultural districts, which was immediately forthcoming, with 5.5 million of the present 8.0 million acres being enrolled by 1978. Controversy continues to surround the basis of the land-use assessment, the impact of revaluations, and the programme's erosion of the local property tax base (Bills and Boisvert, 1987).

California's approach maintained local government control of farmland protection by giving county governments the authority to assist agricultural interests if they wished (Furuseth, 1985). Not surprisingly, the diversity of responsẽs was considerable. Furuseth found, through discriminant analysis, that the profile of a county with a strong farmland protection effort included slow population and housing growth, agricultural orientation, a higher than average tax rate but lower per capita spending, a large black and Hispanic population and a liberal, Democratic-oriented voting tradition.

The Oregon Land Use Act of 1973 is another case of state farmland preservation policies demanding county and local conformity. It took 13 years, however, before the approval of the last county and city plans in 1986 (Nelson, 1990). The implementation of the act resulted in state-local friction. Gustafson *et al.* (1982) noted that 'inherent differences between the forces operating at the state and county levels' essentially prohibited 'an ideal state of collaborative planning'. Local governments both value their autonomy and are sensitive to the influence of special interests in planning. Thus a major research objective for the 1990s should be understanding the dynamics of decision-making in agricultural preservation, as county and municipal governments respond to the introduction or fine-tuning of programmes by state governments.

Federal policies

The Federal government does not intervene directly in the land conversion process in the urban fringe. Federal policies to manage farm output and maintain farm income are strong but barely include the preservation of agricultural land. In the United States, unlike the situation in various European countries, the right to develop land belongs with the land owner, not the national government. The US government is reluctant to control the use of land it does not own. The more local the level, the more acceptable it is to plan land use in the public interest.

The National Land Use Policy Act, proposed in various forms from 1968 to 1975, was the last attempt at comprehensive centralized regulation of privately held lands (Jackson, 1981, p. 61; Popper, 1988, p. 293). Since its narrow defeat in Congress in 1975, the vision of a national land-use plan has receded.

The 1981 Farmland Protection Policies Act merely directs the Department of Agriculture to ensure that federal agencies do not contribute to the loss of agricultural land, and encourages the Soil Conservation Service (SCS) to disseminate information on the Land Evaluation and Site Assessment (LESA) System. LESA helps local governments rate farmlands and directs development away from prime agricultural land. To date more than 400 counties have adopted LESA, but the SCS review of federal agency activities has been rendered tentative by the antiregulation Republican administration in Washington (Bushwick and Hiemstra, 1987). Legislation, such as the 1986 proposal to enable the federal government to cost-share in the purchase of development rights, will continue to be put forward, but little federal involvement in agricultural land-use restructuring can be expected unless government philosophy changes in Washington.

Farmland preservation on the urban fringe and the quiet revolutions in land management

As with federal policy, the preservation of agricultural land on the urban fringe at the state and local level is set within national trends in land-use management. In the late 1960s, American land-use regulation consisted almost entirely of local zoning (Popper, 1988). A 'Quiet Revolution in Land Use' then occurred, so that by 1975 37 states had programmes of statewide planning or review of local decisions. Constantly cited as exemplifying the period are the land-use controls imposed in Oregon, Vermont, California and Florida to protect rural, scenic or coastal areas from development (Healy and Rosenberg: 1979; Jackson, 1981, pp. 69–82; Fulton, 1991).

The conventional view is that the movement lay dormant from 1975 to the late 1980s (Popper, 1988; Fulton, 1991). Popper demonstrates, however, that centralization of regulation at the state level continued, masked by the liberal complaint that it was insufficient, and conservative assertions that it had overextended. Now there is much evidence of a second quiet revolution in land management in the same states, mainly in the Northeast and on the west coast. In the 1990s, however, the movement promises to be national. More traditionally conservative states, such as Texas, Georgia, South Carolina and Arizona, are cautiously exploring growth management, with Georgia, for example, establishing a comprehensive growth management system (DeGrove, 1991, xvii).

While any resurgence of comprehensive land-use management at the state and local level should facilitate the restructuring of agricultural land-use planning programmes on the urban fringe, agricultural preservation could be buried among other issues. New Jersey's adoption of a state planning law in 1986 originated in two State Supreme Court rulings requiring local governments to provide affordable housing, an issue certain to be featured in urban fringe growth management issues of the 1990s (Bushwick and Hiemstra, 1987, p. 197).

Environmental preservation is also certain to be an important goal. Jennings (1989) viewed Popper's article, which won the Journal of the American Planning Association (JAPA) award for the best article published in 1988, as a benchmark for 'just how far the institution of American land-use planning is from coming to grips with the soil, water and biological degradation currently taking place on privately held land.' Popper, however, may have undervalued the environmental focus of recently adopted state land-use programmes (Mandelkar, 1989).

Comprehensiveness is also subject to interpretation. Mandelkar (1989) criticized Oregon's programme, which featured farmland preservation, as comprehensive only in the sense that it thoroughly addressed a narrow range of goals. Conversely, Florida's 1985 Growth Management Act does not explicitly address agricultural land protection. The claim of Bushwick and Heimstra (1987) that this objective would be achieved by the very definition of growth management could be wishful thinking.

When planning on the urban fringe is seen as growth management, the preservation of agricultural land may have no special place in policies. Agriculture may be viewed as competing with active or passive recreation, forests, wetlands and water supply watersheds for space in plans emphasizing housing and other development. As a visually attractive amenity, however, agricultural land has a place. Creative use of cluster development techniques can preserve the visual attributes of farmland and other intrinsically scenic areas from moderate density sprawl (Yaro *et al.*, 1988). In the 1990s, further empirical and theoretical research is needed to maintain the high priority of agriculture in land-use management programmes for the urban fringe.

Enhancing Farm Viability on the Urban Fringe

In the United States, the nature and viability of farming on the urban fringe are less well understood than are the techniques for preserving agricultural land. On the one hand, professional and academic planners are seldom interested in the nature of farming systems; on the other hand the attention of agricultural geographers and other agricultural experts is usually drawn beyond the fringe to areas where 'true' farming exists (Hart, 1991). However, agriculture in the urban fringe has different structural characteristics from those in rural areas. These characteristics are well-researched in Canada (see Chapter 10) and Western Europe (Bryant, 1986), but less is known about the context of the US, especially variations in the characteristics of the urban fringes of different metropolitan areas. Consequently, restructuring the agricultural system in an urban fringe to enhance farm and farmer viability through well-crafted policies is an important research frontier.

The nature of farming on the urban fringe

Recent scholarship in the US (Lawrence, 1988; Nelson, 1990; Hart, 1991) revives von Thünen's model of concentric rings of decreasing land-use intensity around urban areas as a means of understanding the urban fringe. Attention, too, has been paid to Sinclair's (1967) thesis that inverted the rings, so that the least intensity is closest to the city. Hart's own model, empirically derived from the urban fringe of New York City, is consistent with von Thünen's. Every metropolis, he maintains, has a metropolitan bow wave; the outer edge of the urban fringe pushing into rural America like the bow wave of a ship cleaving the sea. It moves incessantly and cannot be halted. The inner edge of the urban fringe likewise inexorably moves outward, signifying the inevitable complete urbanization of land in the fringe. So farming on the fringe is only viable temporarily and farmland preservation is impermanent.

Inside the urban fringe of the New York Metropolitan area, defined as within 50 miles of Times Square, Hart found four bands. On the inner margin next to the fully built-up area was a greenhouse band, then a nursery band, a vegetable band and a dairy band on the outer fringe, which thrusts its bow wave into the truly rural dairy region beyond the fringe. Hart also monitored the progress of these bands through four concentric zones of counties: in 1987 dairying was no longer practised in the outer Zone, IV; vegetables dominated in Zones III and IV; nursery had passed from II into III, but was not yet found in IV; and greenhouses had shifted from Zone I to Zone II, which was now the innermost area of the fringe.

Other scholars are divided over the value of von Thünen's model. Nelson's farmland preservation strategy of an urban growth boundary is based implicitly on von Thünen's principles and the von Thünen landscape he found in Washington County, Oregon (Nelson, 1990). Daniels (1990) asserted, however, that the discontinuous, haphazard pattern on the urban fringe 'renders this model obsolete'. Any prospective agricultural zone on the outer edge of the fringe would be already penetrated by sprawling, decentralized development, occurring first on arterial highways radiating from the central city and then between the arterials, trapping parcels of farmland.

Is the urban fringe better understood in terms of underlying concentric zones of land use or its more apparent heterogeneity? Hart's data aggregating commodity areas and sales at the county level obscure the often haphazard pattern of farms of different sizes combining various categories of product and methods of direct marketing (Rickard, 1991). In addition, berries and tree fruit, which lend themselves to direct marketing methods, do not appear in any of the zones Hart identifies. Questionnaire-based research to test Hart's model also needs to address issues such as the percentage of farmers working off-farm and the amount of idled farmland in the zones.

It is well established that dairying is vulnerable to urbanization. Berry (1979) found on Chicago's urban fringe that dairy farmers were reluctant to invest in buildings and milking equipment because they would be unable to recover this capital if the land were developed. Cash grain farmers, on the other hand, could recover their capital more easily and, therefore, persisted longer as a lower intensity agricultural land use, while urbanization increased around them. This finding was supported by Lawrence's study of seven metropolitan regions between 1949 and 1982 (Lawrence, 1988). Horticulture (including nurseries and greenhouses) proved the most successful adaptation as dairying declined most rapidly in area and value. Low-value field crops proved persistent, whereas poultry did not. The resurgence in production of vegetables, fruit and nuts during the late 1970s, after a gradual decline, may signify new modes of adaptation in urbanizing counties.

Horticulture is clearly an adaptation in an urban fringe, its adoption being one of the ways in which fringe agriculture may become different from that in the truly rural surrounding area. Heimlich (1989) contrasted America's 711 metropolitan counties in the 1985 census with the non-metropolitan ones. Despite the too broad definition of metropolitan, emphasized by the weaker metropolitan characteristics of the 267 counties added to the list since 1970, the census clearly showed the greater intensity of metropolitan agriculture. Despite land conversion, metro counties had the same percentage of land in farms as non-metro counties. The value of products per acre was twice that of non-metro counties. Farms were less than half the size, partly because farmland value was more than double, and also because of the greater percentage of farmers working most of the time off the farm – 40% in metro counties and 32% in non-metro. Only 35% of metro farmers and 40% of non-metro farmers spent no time working off the farm.

Greater intensity was also reflected in the greater use of fertilizer and other chemicals (double) and hired labour (triple) per acre in metro counties. Producing a not inconsiderable 30% of the national value of agricultural products, these counties contributed 70% of the high-value fruit, vegetables and nursery products. High-value crops, in fact, comprised a quarter of the sales of agricultural products in metro areas and a mere 4% in non-metro counties.

Strategies and policies to enhance farm viability

Intensification and off-farm work are evidently adaptive strategies for farmers on the urban fringe, just as increasing either the farmed area or production are the main strategies in rural areas. If conversion to housing and other urban uses threatens farming, it also brings consumers and new marketing opportunities to farmers. A key to sustaining agriculture lies in marketing locally grown products of all kinds (Blobaum, 1987; Vail, 1987). Diversification is a common strategy, with farmers growing a variety of high-value products and marketing them through stores, supermarkets or by various direct means, such as pick-your-own, roadside stands or farmers markets (Rickard, 1991).

Heimlich (1989) argued that new opportunities are emerging for metropolitan agriculture based on consumer preferences for fresh fruit and vegetables, for organically produced food and the increasingly decentralized non-continguous nature of urban sprawl which allows open spaces, farmland and small communities to coexist in the fringe. The blurring of the concepts of rural and urban, he wrote, is exemplified by the increase in the metropolitan area from 10.9% to 16% between 1970 and 1985.

Johnston and Bryant (1987) have usefully summarized the options open to farmers in the urban fringe who wish to make 'positive adaptive' changes to their businesses. They identify: (i) pick-your-own operations, (ii) retail outlets, (iii) land-extensive cash cropping, (iv) direct livestock sales, (v) off-farm employment, and (vi) single-lot severance. The extent to which these strategies are being used in the various urban fringes in the United States represents a research direction for the 1990s.

The degree to which the impermanence syndrome operates and can be meliorated represents another research question. Nelson (1988) found some evidence that for every acre of prime farmland that is urbanized, another half to one acre becomes idled due to farmers' belief that agriculture in that area has no future. A series of negative strategies follows, including disinvestment in farming inputs. Heimlich (1989) however, discovered in the 1985 United States census that the percentage of permanently idled land in metropolitan counties (1.7%) was little higher than that in non-metropolitan counties (1.2%); he also cited evidence of agricultural adaptation to question the importance of the syndrome. Lockeretz *et al.* (1987) found a sample of farmers in the suburbs of Worcester, Massachusetts, to be positive and optimistic about the prospects for farming, with farmers showing an increased commitment to farming. Variation in their expectations and their plans to build up their farms was accounted for more by personal and family factors than by characteristics of the farm or the surrounding area.

Another set of research questions surrounds the issues of part-time farming and pluriactivity in the urban fringe (Smith, 1987; Fuller, 1990). Smith recognized that in the United States middle-size farms are becoming less important than large or small ones. He identified a 'production mode', associated with very large farms, by which expansion of the size of farm achieves economies of scale. Other farmers practise various methods of increasing the value and return to their management. This 'value mode' involves off-farm employment, direct marketing and fewer inputs, which are strategies particularly practised on smaller farms in the urban fringe.

The kind of farming that appears to be most viable on the urban fringe is often held in low regard and ignored in public policy. Hobby farms and part-time farms are usually distinguished pejoratively from highly productive farms (Daniels and Nelson, 1986; Lapping *et al.*, 1989). Federal farm-income policies favour large-scale over small-scale farmers, whereas Daniels (1990) has asserted the need to support middle-income farmers. Another need, however, is to design policies to support the 'value mode' of farm management and positive adaptation strategies. Heimlich asserted that 'agricultural research and extension efforts at both national and local levels could develop programmes aimed at the problems and opportunities facing metro farmers who are making the transition to better-adapted farm types' (Heimlich, 1989).

Another policy and research frontier is the concept of area-wide food system planning at local, county, metropolitan or state scales (Stephens *et al.*, 1988; State of Connecticut, 1989). Potential farm viability on the urban fringe is clearly reduced when the only food system plan in most metropolitan areas is the one informally developed by food wholesalers and retailers (Shortridge and Shortridge, 1989). A metropolitan food plan, according to Blobaum:

> would take into account the quality and availability of close-in farmland, the agricultural system already in place, the amount of food shipped from distant points, the amount of instability existing in growth and development areas, the potential for local sales, the status of the local marketing system, possibilities for extended-season production, and the economic viability of the farm operations involved.
>
> (Blobaum, 1987)

The need for thorough research and constant monitoring of such a food plan is its Achilles heel; otherwise sustaining farmers on the urban fringe and providing fresh food are accepted in principle by planners and consumers (Rickard, 1991). Policies to restructure farming in the urban fringe to create a new sustainable form of agricultural system are as important as devising appropriate policies to preserve the land on which it would take place.

Conclusion

In the end, it is the vision of what the urban fringe should be that will decide policy. There is no federal vision of agriculture on the urban fringe. Whatever restructuring takes place in the 1990s will proceed according to the decisions of state legislatures in cooperation with county and municipal policy makers. The research viewed in this chapter suggests alternatives for these policy makers in the 1990s. Is the urban area to be separated distinctly from the surrounding farming area by as narrow a zone as possible, as is the ideal in Oregon? Should areas of traditional farming, as large as counties, be protected from the advancing patchwork of land conversion by a mixture of prescriptive preservation techniques, including exclusive agricultural zoning, such as might save the Pennsylvania Dutch farmlands? Should the rings of increasingly intensive land use behind the metropolitan bow wave be allowed to eat into America's farmland without hindrance, because there is no good reason to interfere? Should theories of comprehensive growth management be allowed to squeeze farmland out of an urban fringe patchwork of affordable housing, scenic ridges and valleys, wetlands and recreation areas, because flat, prime agricultural land is expensive to preserve and farming with chemicals is ecologically unsound for neighbourhoods? Surely endless streets of single-family homes on small lots will not be the development ideal in the fringe. But will policies to preserve agriculture in the interstices of development and open space be combined with policies to enhance the viability of farming, by assisting farmers to adapt positively to the opportunities of the metropolitan area? Research and experience over the next decade should clarify what policies for agricultural structuring are best suited to the realities and visions of the various urban fringes nationwide.

References

Atash, F. (1987) Urban growth and farmland preservation: an assessment of alternative programs. In: Lockeretz, W. (ed.), *Sustaining Agriculture Near Cities*. Soil and Water Conservation Society, Ankeny, pp. 199–208.

Berry, D. (1979) Sensitivity of dairying to urbanization: a study of northeastern Illinois. *Professional Geographer* 31, 170–6.

Bills, N.L. and Boisvert, R.N. (1987) New York's experience in farmland retention through agricultural districts and use value assessment. In: Lockeretz, W. (ed.), *Sustaining Agriculture Near Cities*. Soil and Water Conservation Society, Ankeny, pp. 231–50.

Blobaum, R. (1987) Farming on the urban fringe: the economic potential of the urban connection. In: Lockeretz, W. (ed.), *Sustaining Agriculture Near Cities*. Soil and Water Conservation Society, Ankeny, pp. 3–8.

Bryant, C.R. (1986) Agriculture and urban development. In: Pacione, M. (ed.), *Progress in Agricultural Geography*. Croom Helm, London, pp. 167–94.

Bushwick, N. and Hiemstra, H. (1987) How states are saving farmland. In: Lockeretz, W. (ed.), *Sustaining Agriculture Near Cities*. Soil and Water Conservation Society, Ankeny, pp. 189–98.

Coughlin, R.E. (1991) Formulating and evaluating agricultural zoning programs. *Journal of the American Planning Association* 57, 183–92.

Daniels, T.L. (1990) Policies to preserve prime farmland in the USA: a comment. *Journal of Rural Studies* 6, 331–6.

Daniels, T.L. (1991) The purchase of development rights: preserving agricultural land and open space. *Journal of the American Planning Association* 57, 421–31.

Daniels, T.L. and Nelson, A.C. (1986) Is Oregon's farmland preservation program working? *Journal of the American Planning Association* 52, 22–32.

DeGrove, J.M. (ed.) (1991) *Balanced Growth: A Planning Guide For Local Governments*. International City Management Association, Washington, DC.

Fuller, A.M. (1990) From part-time farming to pluriactivity: a decade of change in rural Europe. *Journal of Rural Studies* 6, 361–73.

Fulton, W. (1991) The second revolution in land use planning. In: DeGrove, J.M. (ed.), *Balanced Growth: A Planning Guide for Local Governments*. International City Management Association, Washington DC, pp. 116–24.

Furuseth, O.J. (1981) Update on Oregon's agricultural protection program: a land use perspective. *Natural Resources Journal* 21, 57–70.

Furuseth, O.J. (1985) Influences on county farmland protection efforts in California: a discriminant analysis. *The Professional Geographer* 37, 433–51.

Furuseth, O.J. and Pierce, J.T. (1982) *Agricultural Land in an Urban Society*. AAG Resource Publications in Geography, Washington, DC.

Gustafson, G.C., Daniels, T.L. and Shirack, R.P. (1982) The Oregon Land Use Act. *Journal of the American Planning Association* 48, 365–73.

Hart, J.F. (1976) Urban encroachment on rural areas. *The Geographical Review* 66, 1–17.

Hart, J.F. (1991) The perimetropolitan bow wave. *The Geographical Review* 81, 35–51.

Healy, R.G. and Rosenburg, J.S. (1979) *Land Use and the States* 2nd edn. The Johns Hopkins University Press, Baltimore.

Heimlich, R.E. (1989) Metropolitan agriculture: farming in the city's shadows. *Journal of the American Planning Association* 55, 457–66.

Jackson, R.H. (1981) *Land Use in America*. John Wiley, New York.

Jennings, M.D. (1989) The weak link in land use planning. *Journal of the American Planning Association* 55, 206–8.

Johnston, T.R.R. and Bryant, C.R. (1987) Agricultural adaptation: the prospects for sustaining agriculture near cities. In: Lockeretz, W. (ed.), *Sustaining Agriculture Near Cities*. Soil and Water Conservation Society, Ankeny, pp. 9–21.

Lapping, M.B., Daniels, T.L. and Keller, J.W. (1989) *Rural Planning and Development in the United States*. The Guilford Press, New York.

Lawrence, W. (1988) Changes in agricultural production in metropolitan areas, *The Professional Geographer* 40, 159–75.

Lockeretz, W., Freedgood, J. and Coon, K. (1987) Farmers' views of the prospects for agriculture in a metropolitan area. *Agricultural Systems* 23, 43–61.

Mandelkar, D.R. (1989) The quiet revolution: success and failure. *Journal of the American Planning Association* 55, 204–8.

Nelson, A.C. (1987) How regional planning influences rural land values. In: Lockeretz, W. (ed.), *Sustaining Agriculture Near Cities*. Soil and Water Conservation Society, Ankeny, pp. 263–76.

Nelson, A.C. (1988) An empirical note on how regional urban containment policy influences an interaction between greenbelt and exurban land markets. *Journal of the American Planning Association* 54, 178–84.

Nelson, A.C. (1990) Economic critique of US prime farmland preservation policies. *Journal of Rural Studies* 6, 119–42.

Pease, J.R. (1991) Farm size and land-use policy: an Oregon case study. *Environmental Management* 15, 337–48.

Platt, R.H. (1985) The farmland conversion debate. *Professional Geographer* 37, 433–42.

Popper, F.J. (1988) Understanding American land use regulation since 1970: a revisionist interpretation. *Journal of the American Planning Association* 54, 291–301.

Rickard, T.J. (1991) Direct marketing as agricultural adaptation in Megalopolitan Connecticut. In: van Oort, G.M., van den Berg, L.M., Groenendijk, J.G. and Kempers, A.H. (eds), *Limits to Rural Land Use*. Pudoc, Wageningen, pp. 79–88.

Rose, J.B. (1984) Farmland preservation policy and programs. *Natural Resources Journal* 24, 591–640.

Shortridge, B.G. and Shortridge, J.R. (1989) Consumption of fresh produce in the metropolitan United States. *The Geographical Review* 79, 79–98.

Sinclair, R. (1967) Von Thünen and urban sprawl. *Annals of the Association of American Geographers* 57, 72–87.

Smith, S.N. (1987), Farming near cities in a bimodal agriculture. In: Lockeretz, W. (ed.), *Sustaining Agriculture Near Cities*. Soil and Water Conservation Society, Ankeny, pp. 77–90.

State of Connecticut Department of Agriculture (1989) *Connecticut Grown: 1989 Annual Report*. Department of Agriculture, Hartford.

Stephens, G.R. Fleming, J.G., Gacoin, L. and Bravo-Ureta, B. (1988) *Better Nutrition in Connecticut: Opportunities for Expanding Fresh Produce Production and Consumption*. Bulletin 852, The Connecticut Agricultural Experiment Station, New Haven.

Vail, D. (1987) Suburbanization of the countryside and the revitalization of small farms. In: Lockeretz, W. (ed.), *Sustaining Agriculture Near Cities*. Soil and Water Conservation Society, Ankeny, pp. 23–36.

Yaro, R.D., Arendt, R.G., Dodson, H.L. and Brabec, E.A. (1988) *Dealing With Change in the Connecticut River Valley: A Design Manual for Conservation and Development.* Center for Rural Massachusetts, University of Massachusetts, Amherst.

13

Forestry as an Alternative Land Use: A British Perspective

Charles Watkins

The 1980s saw some dramatic changes in British forestry policy. There was a questioning of the assumptions that have lain behind forest policy for much of the present century. This questioning was concerned with a wide range of issues including the economic backcloth to forestry, the ecological effects of afforestation and of different types of woodland management, the different uses to which woodland should be put, and woodland as an alternative use of agricultural land. This chapter summarizes some of the principal changes and then concentrates on two main themes: the growing recognition of the importance of ancient woodland and the increasing emphasis on lowland afforestation.

British Woodland

The area of woodland in Britain has roughly doubled over the present century and now covers about 10% of the total land area. Compared with other members of the EC, the US and Canada, however, the proportion of woodland is still low. European countries of a comparable size to Britain, such as France, Germany and Italy, are all well over a fifth wooded (see Table 13.1). Within Britain, woodland is distributed very unevenly. The most heavily wooded areas include parts of Scotland, such as Galloway, and parts of southern England, such as Hampshire and Sussex. The types of woodland found in these areas is, however, dramatically different. In Galloway, most of the woodland consists of recent coniferous plantations whereas in Hampshire and Sussex, most of the woodland is broadleaved and much is ancient.

The doubling of the British woodland area in the 20th century is due largely to the government policy of woodland expansion which has its roots in a series of late 19th century government reports, but which was essentially established with the formation of the Forestry Commission in

Table 13.1. Area of woodland in selected European and North American Countries.

Country	Woodland (%)
Britain	10
France	27
West Germany	30
Italy	23
Spain	31
USA	29
Canada	39

Source: *Forestry Facts and Figures, 1989–1990.* Forestry Commission (1991). All figures apart from Britain (1991) relate to 1989.

1919. The long-term aim was to establish a suitable strategic reserve of timber. The target area of 2 million ha has now been reached and in the 1980s there has been a general questioning of the assumptions underlying previous forestry strategy (Mather, 1991). A number of reports have been produced that challenge the conventional economic assumptions underlying commercial afforestation (for example, National Audit Office, 1986).

In addition to economic critiques, the long-standing arguments that coniferous afforestation in the upland areas was causing significant problems in terms of reducing the nature conservation value of the planted land, reducing the quality and extent of public access on hill land and damaging the scenic quality of the uplands came to a head (Tompkins, 1989). One key issue was the battle over the afforestation of parts of the Flow country of Caithness and Sutherland (Stroud *et al.*, 1987; Nature Conservancy Council, 1988). A major policy change has been the recent decision to stop the income tax concessions relating to forestry management, which had been in operation since 1916.

Although the forestry industry can be seen as being in need of new justifications for expansion, there is still an enthusiastic forestry sector (Forestry Industry Committee of Great Britain, 1987, 1990). General concern about agricultural surpluses in the EC has prompted consideration of forestry as an alternative land use. The need to withdraw land from the utilized agricultural area of the EC (Lee, 1991) has led to a policy shift away from planting in the English uplands (other than re-stocking), towards lowland afforestation, including schemes such as the MAFF Farm Woodland Scheme (see Chapter 7), the new National Forest in the Midlands, and a series of community forests around large towns and cities. The shift in interest to lowland woodland management is also reflected in the Forestry Commission's broadleaved policy (Watkins, 1986). In addition, there is greater interest in timber as a farm crop. This is derived from a desire to diversify the farm economic base and a concern for high-quality timber to offset the generally poor-quality timber associated with upland plantations. These aims could be achieved by developing farm-forestry and agro-forestry systems (Renewable Energy Enquiries Bureau, 1991).

The 1980s have seen many of the forests established in the 1920s and 1930s reaching maturity. There is now increasing interest in imaginative ways of ensuring that the second rotation takes full account of conservation, landscape and recreational objectives. There has also been a massive increase in interest in the wildlife conservation importance of woodlands of all sorts. Perhaps the most interesting development has been the increasing recognition given to the concept of ancient woodland. This was only recognized by a limited number of specialists in the late 1970s, but has since become well established and is now the basis of important conservation policies (Rackham, 1980; Peterken, 1981).

Ancient Woodland

Over the last 20 years our knowledge of the history and ecology of British woodland has been transformed. It is now widely thought that many woods throughout the country are the descendants of the primeval woodland that covered the country before humans started to clear it for agriculture, or manage it for wood and timber. This new interest in woodland history has stemmed from the work of historical ecologists. Documentary sources and field evidence have been brought together to show that many woods are of great antiquity (Salbitano, 1989; Watkins, 1989, 1990).

The original woodland cover may be termed natural woodland because it was largely unaffected by human activity. No natural woodland exists in Britain today because all woodland has either been cleared and destroyed, or managed in some way. Woods that have survived clearance to the present day, and which are remnants of the natural woodland are called primary. Although they have been affected by human activity, such as the grazing of domesticated animals or coppicing, the sites on which they grow have never been ploughed or cultivated.

The long-term trend until the 18th century was woodland clearance, but there were periods when land already cleared became wooded again. This new woodland, growing on formerly open ground, is known as secondary woodland. It results from the reversion of land to woodland through natural regeneration or the making of plantations. Natural succession to woodland tends to take place when land falls out of cultivation or when the grazing pressure is reduced. This process occurs in periods of agricultural depression, such as in the late 19th century. In the present century new secondary woodland has sprung up on abandoned commons, awkward corners in fields, old quarries, disused railway lines and ungrazed and uncultivated chalk downs.

Most modern secondary woodland has been specifically established in the form of plantations. Landowners from the late 17th century onwards made many new plantations. Plantations were an essential element of the parks formed by landscapers such as Brown and Repton, and in addition to larger commercial plantations many small plantings were established as fox and game coverts (Daniels, 1988; Daniels and Watkins, 1991). Most secondary woodland today, however, consists of the extensive plantations made by the Forestry Commission and the private forestry companies in the uplands and on the lowland heaths since the First World War.

It is useful to distinguish between ancient woods, which originated in or before the 16th century, and recent woods, which originated from the 17th century onwards (Rackham, 1980; Peterken, 1981). Although many plantations were made in the 18th and 19th centuries, little ancient woodland was planted-up until the present century. By 1900, the market for coppice products had almost collapsed and during the first half of the present century coppices became increasingly neglected. At the same time, a national forest policy which depended largely on the planting of even-aged plantations for the rapid growth of timber became established. The way was clear, therefore, for the conversion of much coppice into plantation.

Once the conservation importance of ancient woods became generally recognized (Rackham, 1976), steps began to be taken to protect them and to ensure that they were managed sympathetically. The House of Lords, in a report on forestry (Sherfield, 1980), recommended separate measures to conserve the surviving ancient woods, and the Nature Conservancy Council (NCC) (English Nature since 1991) elaborated these proposals (Steele and Peterken, 1982). They considered that new afforestation and the majority of existing recent woodland would continue to be intensively managed as plantations, but that ancient woods should be retained largely as sources of hardwood timber which should be managed to conserve their nature and landscape conservation value.

How do the existing state controls over woodland management, both voluntary and compulsory, come into play with ancient woodland? The situation is complicated. First, the NCC laid down detailed criteria by which ancient woodland could be defined. Second, the various woodland grant schemes have been modified to persuade owners not to replant ancient woodland with trees that are likely to damage the nature conservation value of the sites. Third, whether a wood is classed as ancient or not has become an important consideration in the making of woodland tree preservation orders and the granting of felling licences. These different approaches are now discussed in greater detail.

Ancient woodland inventory

The NCC considered that special measures could only be applied to ancient woods if they could be clearly identified. Consequently, an inventory of all ancient woodlands over 2 ha in area was instigated and this has recently been completed on a county basis for the whole of Great Britain (Kirby *et al.*, 1984; Spencer and Kirby, 1992). This inventory has in practice been a massive exercise in applied historical geography; it has made use of documentary sources, such as early editions of ordnance survey maps and estate plans, in conjunction with some ecological survey data. The NCC sees these inventories as having a number of key functions. These include:

1. a basis for monitoring future changes in the number and extent of ancient woods, and indicating the extent of recent losses;
2. a basis for consultations about any forestry practice that may be applied to ancient woodland, and for the advice NCC may give about individual sites;
3. a basis for selecting Sites of Special Scientific Interest (SSSIs).

The initial results of the ancient woodland inventory indicate that since 1945 the 500 000 ha of surviving ancient, semi-natural woodland in Britain has been greatly reduced: 10% has been destroyed by clearance, mostly for agriculture, and 30% has been converted into plantations. Half of the total area has stood largely unmanaged since 1945, and only 10% has survived under traditional forms of management. Currently around a third of all woodland in England and Wales is classed as ancient, and a fifth is ancient semi-natural woodland.

Modifications of grant schemes to take account of ancient woodland

Following wide-ranging discussions, the Forestry Commission introduced a broadleaves policy in 1985 (Watkins, 1986). A new Woodland Grant Scheme was introduced in 1988; this included special provisions for ancient woodland. Anyone applying for one of the grants had to inform the Forestry Commission whether any of the woodland they intended to plant was classed as ancient. The Forestry Commission noted that the entries in the NCC's inventory of ancient woodland were 'subject to verification and any dispute about validity should be stated' at the time the grant application was made. Special provisions were made for the ancient native pinewoods of Scotland (Forestry Commission, 1988). The Forestry Commission's policy was that woods included on the inventory should be managed with due regard to their intrinsic conservation and historical interest.

In 1991 a new Woodland Grant Scheme was introduced (Forestry

Commission, 1991) which further emphasizes the shift in funding towards the management of ancient woodland. The scheme introduces grants to encourage woodland management, while special management grants are available for woods of special environmental value, provided that suitable management to maintain and enhance the character of this woodland is carried out. The Forestry Commission state that there is a presumption that woods classed as ancient and semi-natural on the inventory drawn up by the NCC will qualify for the special management grants of £45 per hectare in woods of less than 10 ha, which is three times the amount that will be paid for the management of coniferous woodland.

Compulsory woodland controls

Landowners who wish to replant ancient woodland without the assistance of grants are not affected by the voluntary controls discussed so far. They are affected, however, by the need to apply for felling licences (Griffin and Watkins, 1988). Current policy is not to allow the replacement of broadleaved trees with conifers. Certainly, the Forestry Commission will no longer allow the felling of ancient semi-natural woodland and its replacement with conifers. There is, however, a weakness in the felling licence system: it is possible to fell whole coppice woods, many of which are ancient semi-natural stands, so long as the coppice is below 15 cm in diameter, without the need to apply for a licence. Tree preservation orders (TPO) may, however, be used to protect ancient coppice woods from indiscriminate clearance. A report published by Hampshire County Council in 1988 noted that experience in 'Hampshire indicates that [TPOs] are still necessary to prevent the clearance of coppice woodland composed mainly of material below licensable size, where the Forestry Commission is powerless to act'. The report goes on to note that at least three cases had occurred in Hampshire in 1987–1988 that confirmed the value of keeping TPOs on woodland.

However, TPOs also have their weaknesses. They are designed to protect trees; they do not protect associated habitat such as the ground flora, shrub layer and soils. Thus it is possible for owners of woodland with a low density of mature trees legally to cultivate between the scatter of trees and convert the woodland to parkland. The trees would survive, but all the associated characteristics of woodland important in amenity and conservation terms would be destroyed. Moreover, under the Town and Country Planning (Amendment) Act 1985, trees within woodland TPOs have to be replaced if they have been removed, uprooted or destroyed, by planting the same number of trees either on or near the land on which the original trees stood. However, the act allows for a considerable amount of flexibility and compromise between the local authority and the owner. Not only is there room for discussion as to the appropriate size and species of the replacement trees, but the site of the replacement trees can be changed. The site may be agreed between the two parties, or designated by the local authority. It is important that local authorities are aware that, in the case of ancient semi-natural woodland, the flora and soils of the woodland *site* itself can be of as great ecological and historical importance as the trees growing on it.

The New Lowland Forests

Until recently the idea of establishing large areas of new woodland in lowland Britain would have been treated with surprise and with considerable caution. The current surplus of agricultural land in the EC has, however, made such a proposal a realistic option. The Countryside Commission and Forestry Commission schemes for new community forests around urban areas (Countryside Commission, 1989a), together with the separate proposal for a new national forest in the Midlands (Countryside Commission, 1989b), have become an accepted part of the national programme of land-use change.

Estimates of the extent of surplus agricultural land in the UK vary from a sixteenth to just over a fifth of the total area of agricultural land. This could mean anything from 1 million to 5 million ha of surplus agricultural land by 2015 (House of Commons, 1990, p. ix). Various ways of reducing the acreage of arable land in the UK have been proposed and introduced. These include the setting aside of agricultural land as fallow, and the use of land for 'soft' non-agricultural uses such as golf courses and country parks. The House of Commons Agriculture Committee's Report on *Land Use and Forestry* suggests that these types of land-use change are only likely to use up around half the surplus land available. They conclude that the 'significant alternative land use in the next 20 years is likely to be forestry' and that

> planting targets since the War have aimed at a modest expansion in the country's forestry estate: there now seems to be the scope, if not the necessity, for a far greater emphasis on the role of woodlands and forestry in the process of rural development.
>
> (House of Commons, 1990, p. xv)

The great significance of this surplus agricultural land is that it allows afforestation to be brought 'down the hill' from the uplands to the lowlands. Upland areas contain the largest areas of semi-natural vegetation remaining in Britain; areas that are especially important from the point of view of nature conservation because they have not been fragmented by agricultural intensification as in most of lowland Britain. The current surplus of agricultural land means that one of the principal locational factors affecting afforestation throughout the 20th century, namely that afforestation should only take place on 'unimproved' land of low agricultural value, is no longer of paramount importance.

Potential uses of new lowland forests

Woodland can be used for a very wide range of different purposes. The most important uses in Britain are timber production (whether for commercial purposes or for use as fencing material or firewood by the owner), landscape, nature conservation, recreation and public access, game conservation and shelter. An important aspect of woodland management is the way in which these different uses may be combined in different ways depending on the management objectives of the owner (Watkins, 1987).

Within the new forests it is likely that the different woods will have different management objectives depending on the management aims of the owners. The new forests as a whole will be used for a multiplicity of purposes, but individual woods will tend to be used for specific purposes. Some woods will be used principally for timber production, although recreation, landscape value or nature conservation may be additional management aims. Other woods may be managed principally for recreation, landscape or nature conservation, although these will frequently provide some income from timber production.

There are several potential silvicultural benefits arising from the new lowland woodlands (Watkins, 1991). The new forests will be able to take advantage of a wide range of commercial species, both broadleaved and coniferous. In the uplands, the range is generally limited to Sitka spruce, Lodgepole pine and the larches. In the lowlands, greater emphasis could be placed on Douglas fir, Corsican pine and broadleaved species; there is also the opportunity to increase the quality of timber coming onto the market.

In lowland forestry a greater range of silvicultural systems is applicable than in the uplands and the use of selection and shelterwood systems is more feasible. This will tend to result in higher costs, but thinning is likely to be more practicable and there is a greater likelihood that an earlier and more positive cash flow can be achieved. Although the recent great storms of 1987 and 1990 have demonstrated that woodlands in all parts of the UK are subject to storm damage, the risk of windthrow is in general considerably less in the lowlands than in the uplands.

Finally, if the new areas of woodland are established on 'improved' agricultural areas, there is little need for specialist forestry ploughing; indeed ploughing of any sort would frequently be unnecessary where the establishment of trees followed the harvest of an arable crop. Most arable land has been drained, where necessary, in the past. An additional benefit is that owing to the high quality of the land no fertilizers will be required, although this high quality may also mean that there are relatively high levels of competition from weed growth and weeding costs will therefore be higher. Establishment costs may be reduced by the lesser need to provide expensive deer fencing compared with certain upland areas, but, with the currently expanding deer population in the lowlands, this advantage may be relatively unimportant. Moreover, formidable forest pests, such as the rabbit and the grey squirrel, are likely to cause considerable problems in the new lowland woods.

English Nature (previously the NCC) has been quick to see the potential nature conservation benefits of this policy shift. They have argued that the afforestation of large areas of lowland agricultural land should, at the national scale, reduce the pressure to afforest land of nature conservation importance, and that the new planting should contribute substantially to the national planting target of 33 000 ha. In practice, however, it appears likely that there will be continued pressure for upland afforestation. However, the increased scope for new commercial woodlands in the lowlands may reduce the pressure to manage ancient woodland in a manner unsympathetic to the nature conservation interest, especially when the newly introduced management grants are taken into account. The establishment of woodland on arable and improved grassland will usually result in a net gain in terms of nature conservation. Coniferous plantations provide better wildlife habitat than intensively farmed land. There will also be opportunities to create habitats of nature conservation interest on derelict industrial sites.

The new lowland woods and the associated open land will also provide increased opportunities for woodland recreation. This will in turn reduce the recreational pressure on existing semi-natural habitats of high conservation value, such as ancient woodland and lowland heaths. Moreover, greater access to woodland could provide ample opportunity to explore, develop and demonstrate the merits of the multiple use of woodlands and associated open land.

Planning and implementation of the new forests

During the 1980s the Forestry Commission's enterprise role became more concerned with the re-stocking of existing woodlands than the planting of bare ground. This has led to an increasing acceptance of the case for enhanced control both over the allocation and design of future afforestation (Brotherton, 1986). The Countryside Commission called for planning controls, the Countryside Commission for Scotland (who experienced the greatest pressure for afforestation) argued only for planting licences, whereas the Convention of Scottish Local Authorities championed Indicative Forestry Strategies. These have now been accepted at an official level in Scotland and on a more informal basis by English and Welsh counties.

Forestry and woodland strategies are documents, usually drawn up by the local authority, that identify those areas that are suitable for different types of forestry and those areas that are best left unplanted. These strategies enable foresters to take account of sensitive areas when making planting proposals and identify areas where they are unlikely to encounter conflict. They could be a useful means of reducing disputes over the allocation of different land uses, and of bringing different interests together. However, to succeed they need 'to incorporate the full range' of land-use interests (House of Commons, 1990, p. xxii).

There may well be considerable difficulties in encouraging landowners to establish new areas of woodland. Indeed, one of the conclusions of a recent research project is that existing grant structures appear to be 'insufficient to implement the concept of Community Forests' (Bishop, 1990, p. 402). This project included surveys of farmers, property developers and mineral companies. The results of these surveys indicated that, on farmland, planting was likely to be limited and would tend to be associated with 'hobby' farmers and farms where game management was an important consideration.

Property developers may carry out small-scale tree planting in advance of development in order to increase their chances of obtaining planning permission, or as part of a planning gain package, especially if the development was associated with leisure. However, all types of developer considered afforestation of their land holdings to be unrealistic. They were concerned about costs and problems of management, and considered woodland establishment would decrease the chance of obtaining planning permission in the future. A general conclusion that can be drawn from the results of this and other studies is that the levels of grant payable to landowners who establish new areas of woodland will probably have to be increased considerably if substantial areas of land are to be converted into woodland.

One general problem identified by the Bishop study is that the establishment of new areas of woodland is frequently made more difficult by the increasing use of new forms of tenure. Within agriculture there are now many types of short-term tenancies and these are increasing the already diverse nature of land occupation (Ward *et al.*, 1990; Winter *et al.*, 1990). In addition, much land, although owner-occupied, is covered by options and conditional contracts with developers and mineral companies. Bishop considers that the effect of these legal factors is to 'increase land use uncertainty' and restrict the 'chances for multi-purpose woodland creation . . .' (1990, p. 409).

One way of encouraging the establishment of new areas of woodland might be through the use of 'planning gain'. There are a number of statutory mechanisms that enable benefits through the planning system to be secured by a local authority. These mechanisms could be used to establish new woodland or in securing provisions for public access or nature conservation within lowland forestry. There is little practical experience of the use of such mechanisms in this context, however, and there is evidence that they should be used with great care as the practice can undermine or distort the planning system.

Conclusion

The long-term nature of woodland management and forestry means that they are peculiarly susceptible to changes in government policy; the past decade has not been unusual in that it has witnessed substantial shifts in attitudes to woodland management in the UK. At the same time, although it has fluctuated due to market conditions and government incentives, such as grants and tax concessions, the pressure to afforest large areas of the uplands, especially in Scotland, remains. It is currently government policy to afforest 33 000 ha a year and to establish a further 12 000 ha annually under the Farm Woodland Scheme. This chapter has described how two shifts of attitude – the recognition of the importance of ancient woodland and the move towards lowland afforestation – have developed and how state grants and controls have been modified to take account of changing priorities. In general, forestry policy continues to be modified very much on an *ad hoc* basis. Although forestry attracts high levels of public sector financial support, there is still no coherent national strategy for the planning and management of forestry land.

References

Bishop, K.D. (1990) Multi-purpose woodlands in the countryside around towns: fact or fiction? Unpublished PhD thesis, University of Reading.

Brotherton, I. (1986) Agricultural and afforestation controls: conservation and ideology. *Land Use Policy* 3, 21–30.

Brouwer, F.M. and Chadwick, M.J. (1991) Future land use patterns in Europe. In: Brouwer, F.M., Thomas, A.J. and Chadwick, M.J. (eds), *Land Use Changes in Europe*. Kluwer Academic, Dordrecht, pp. 1–20.

Countryside Commission (1989a) *Forests for the Community*. Countryside Commission, Cheltenham.

Countryside Commission (1989b) *A New National Forest in the Midlands: A Consultation Document*. Countryside Commission, Cheltenham.

Daniels, S. (1988) The political iconography of woodland in later Georgian England. In: Cosgrove, D. and Daniels, S. (eds), *The Iconography of Landscape*. Cambridge University Press, Cambridge, pp. 43–82.

Daniels, S. and Watkins, C. (1991) Picturesque landscaping and estate management: Uvedale Price at Foxley, 1770–1829. *Rural History* 2, 141–70.

Forestry Commission (1988) *Woodland Grant Scheme*. Forestry Commission, Edinburgh.

Forestry Commission (1991) *Grants and Procedures*. Forestry Commission, Edinburgh.

Forestry Industry Committee of Great Britain (1987) *Beyond 2000: the Forestry Industry of Great Britain*. FICGB, London.

Forestry Industry Committee of Great Britain (1990) *Options for British Forestry 1989–1990*. FICGB, London.

Griffin, N. and Watkins, C. (1988) The control of tree felling: recent developments in statute and case law. *Quarterly Journal of Forestry* 82, 26–32.

Hampshire County Council (1988) *The Future of Hampshire's Existing Woodland*. Hampshire County Council, Winchester.

House of Commons (1990) Agricultural Committee, 2nd Report. *Land Use and Forestry* Vol. 1 (Session 1989–1990). HMSO, London.

Kirby, K.J., Peterken, G.F., Spencer, J.W. and Walker, G.J. (1984) *Inventories of Semi-natural Woodland*. Nature Conservancy Council, Peterborough.

Lee, J. (1991) Land resources, land use and projected land availability for alternative uses in the EC. In: Brouwer, F.M., Thomas, A.J. and Chadwick, M.J. (eds), *Land Use Changes in Europe*. Kluwer Academic, Dordrecht, pp. 1–20.

Mather, A.S. (1991) Pressures on British forest policy: prologues to the post-industrial forest. *Area* 23, 245–53.

National Audit Office (1986) *Review of Forestry Commission Objectives and Achievements*. HMSO, London.

Nature Conservancy Council (1988) *The Flow Country: The Peatlands of Caithness and Sutherland*. Nature Conservancy Council, Peterborough.

Peterken, G.F. (1981) *Woodland Conservation and Management*. Chapman and Hall, London.

Rackham, O. (1976) *Trees and Woodland in the British Landscape*. Dent, London.

Rackham, O. (1980) *Ancient Woodland*. Edward Arnold, London.

Renewable Energy Enquiries Bureau (REEB) (1991) *Arable Coppice*. REEB, Harwell.

Salbitano, F. (ed.) (1989) *Human Influence on Forest Ecosystems Development in Europe*. Pitagora Editrice Bologna, Bologna.

Selman, P. (1990) Forestry and land use planning: a case for indicative strategies. *Arboricultural Journal* 14, 53–9.

Sherfield, Lord (1980) *Scientific Aspects of Forestry*, 2nd report of the House of Lords Select Committee on Science and Technology. HMSO, London.

Spencer, J.W. and Kirby, K.J. (1992) An inventory of ancient woodland for England and Wales. *Biological Conservation* 62, 77–94.

Steele, R.C. and Peterken, G.P. (1982) Management objectives for broadleaved woodland conservation. In: Malcolm, D.C., Evans, J. and Edwards, P.N. (eds), *Broadleaves in Britain*. Institute of Chartered Foresters, Edinburgh, pp. 91–103.

Stroud, D., Reed, T., Pienkowski, M. and Lindsay, R. (1987) *Birds, Bogs and Forestry*. Nature Conservancy Council, Peterborough.

Tompkins, S. (1989) *Forestry in Crisis: the Battle for the Hills*. Christopher Helm, London.

Ward, N., Marsden, T. and Munton, R. (1990) Farm landscape change: trends in upland and lowland England. *Land Use Policy* 7, 291–302.

Watkins, C. (1986) Recent changes in government policy towards broadleaved woodland. *Area* 18, 117–22.

Watkins, C. (1987) The future of woodlands in the rural landscape. In: Lockhart, D. and Ilbery, B. (eds), *The Future of the British Rural Landscape*. Geo Books, Norwich, pp. 71–96.

Watkins, C. (1989) The idea of ancient woodland in Britain from 1800. In: Salbitano, F. (ed.), *Human Influence on Forest Ecosystems Development in Europe*. Pitagora Editrice Bologna, Bologna, pp. 237–6.

Watkins, C. (1990) *Woodland Management and Conservation*. David and Charles, Newton Abbot.

Watkins, C. (1991) *Nature Conservation and the New Lowland Forests*. Nature Conservancy Council, Peterborough.

Winter, M., Richardson, C., Short, C. and Watkins, C. (1990) *Agricultural Land Tenure in England and Wales*. Royal Institution of Chartered Surveyors, London.

14

RESEARCH AS AN ALTERNATIVE LAND USE

Lisa Harrington

Land areas controlled by public agencies and by non-governmental organizations in the United States and elsewhere often are explicitly devoted to research use, or to a combination of uses including research. This has followed from both practical needs and social emphases common in the development of economically advanced countries. Regions where research has gained fairly high levels of importance as a designated use or purpose of particular lands are usually quite rural, as opposed to rural/urban fringe areas. Some research areas, however, have developed in fringe areas or, much more rarely, have developed into relatively urbanized areas simply because of the volume of research activities taking place. Los Alamos, New Mexico, would be an example of the latter case. This chapter constitutes a review of the development of rural land for research purposes, focusing on the case of the United States.

Development of Research Lands

Rural research lands may be categorized in a number of ways, depending on the major purposes(s) of specific areas and/or the agencies or groups operating them. A major delineation is made between experimental and non-experimental ('observational' or natural) research areas. A major difference exists here between approaches to research use of designated areas; a further disparity exists between experimental research areas dependent upon the purposes and types of experimental research pursued.

Experimental research areas

Basic governmental and societal philosophies or outlooks toward the roles of science and research have shifted noticeably during the last century. In the late 1800s, the United States government initiated the process of designating lands for research with the 1887 passage of the Hatch Act, following the Land Grant Act of 1862. The Land Grant Act donated land for state and territorial agricultural colleges, and the Hatch Act created the system of state Agricultural Experiment Stations. The purpose of the Agricultural Experiment Stations was 'to aid in acquiring and diffusing among the people of the United States useful and practical information on subjects connected with agriculture, and to promote scientific investigation and experiment[ation] respecting the principles and applications of agricultural science' (Hatch Act, section 1).

During the late 1800s the United States was still primarily an agrarian society. Further development and the settlement of vast western regions, in particular, were unquestioningly seen to be closely tied to agriculture. Thus, it could be argued that the federal government, through its 'donations' and directives to the states, was attempting to meet practical economic needs. Dramatic erosion problems also contributed to the need for some type of action (Schlesinger, 1983, p. 363). Today, in addition to the state agricultural experiment stations, there are federally operated experimental reserves tied to the United States Department of Agriculture (USDA). These include USDA Experimental Forests, Experimental Rangelands, and Experimental Watersheds. Agriculturally oriented experimental research areas operated at the state or federal level vary greatly in size, from less than 5 ha to tens of thousands of hectares. (Jornada Experimental Range in New Mexico, for example, is over 78 000 ha in size.) In addition, there are numerous agricultural research sites that have been established in recent years by private industry. These include companies specializing in various agricultural technologies, such as hybrid development, artificial insemination and agricultural chemicals. Agriculturally oriented research areas are certainly not limited to the United States, although in countries such as the United Kingdom their total spatial extent obviously will be more limited.

The establishment of experimental research areas is often linked with technological advances and economic or strategic concerns. Experimental research includes the development and testing of various technologies. Although early experimental research sites in the United States were primarily focused on the expansion of understanding of proper agricultural land resource management and the development of agricultural technologies, the past half-century of expansion of rural experimental research areas has involved much greater emphasis on other aspects of society. With increasing national concerns regarding military security and international economic competition, rural areas have been transformed by the establishment of experimental research areas designed to achieve and maintain military superiority or obtain high levels of energy availabilty. Experimental research areas established since the early 1940s have largely centred on (i) the military, and (ii) energy development (especially nuclear).

The two are often linked and share the characteristic of involving hazardous materials and/or activities. Given the uncertainties involved in this research and desires for security, isolated (extremely rural) sites were selected for many of these areas. This occasionally involved the application of 'eminent domain' and the displacement of agriculturalists (for example, at White Sands, New Mexico).

Some military installations are logically included as experimental reserves; examples include White Sands Missile Range in New Mexico and the Desert Test Center/Dugway Proving Grounds/Wendover Range complex in Utah. White Sands Missile Range is one of the largest research areas in the world, with well over 780 000 ha. In spite of its name, White Sands is a research and testing facility with a variety of projects, and has been characterized as 'just one huge laboratory' (Eckles, 1989, personal communication). Research projects at White Sands have included climate and environmental tests, simulations, clothing tests and missile tests (Eckles, 1989, personal communication). Trinity Site, where the first atom-bomb detonation took place, is located on the installation.

Several large United States Department of Energy (DOE) reserves are also primarily research-oriented. Hanford (Nuclear) Reservation in Washington State, the Nevada Test Site, and the Idaho National Engineering Laboratory are in this group, as well as more modest, and perhaps more benign, sites like the Southwest Region Solar Experiment Station in Las Cruces, New Mexico. Another large area is Los Alamos National Laboratory, which covers approximately 11 250 ha. It was established during war-time (to build the atomic bomb) in the early 1940s. Research activities here continue to include nuclear weapons design and testing, but a variety of other research topics in archaeology, biology, geothermal energy, chemistry and physics also are pursued. The laboratory is one of five designated DOE National Environmental Research Parks (Bildstein and Brisbin, 1990); it currently employs about 8000 persons (Schwartz, 1989, personal communication) and its fiscal year 1985 budget was US$915 million (Stephens, 1989, personal communication). Los Alamos County can be considered a research county; it was created from parts of adjacent counties in response to the establishment and growth of the laboratory. The establishment of this research area has resulted in a kind of 'suburbanization' of the locale, although it is located in a rugged mountainous and plateau region that is generally rural and could have been expected to remain entirely rural without the establishment of the laboratory.

Although most DOE sites were established for the development of energy-related technologies, the waste materials associated with a variety of modern activities has led to fairly recent recognition of the need for research into proper storage and disposal of hazardous materials. The Waste Isolation Pilot Plant in southeastern New Mexico, for example, was established in the 1980s as a research site while a possible low-level nuclear waste disposal facility is developed. The DOE is now facing a cleanup of many of its sites, where hazardous waste materials were improperly handled in the past.

Natural research areas

During the late 1800s there was increasing public and governmental concern for the conservation/preservation of natural areas, as well as with agricultural management. At that time, the major concern simply was to save outstanding examples of 'wild' nature for recreational use and to fulfil a moral responsibility. The understanding then was that to designate an area as off-limits to particular uses was to protect the area and the natural processes at work there. This perception has been deeply entrenched in modern societies' approaches to nature preservation; most people see an oxymoron in 'wilderness management'. Only comparatively recently has our comprehension of environmental interconnections led to increased attempts to study the things we wish to maintain, in order to better protect them and to understand both natural and human-induced environmental changes.

Preservation in the United States, Australia and Canada started on a fairly large scale in the late 1800s, with the first National Parks at Yellowstone (1872), Royal (1879) and Banff (1885). Other types of areas, such as United States Forest Reserves (now National Forests) and National Monuments, followed. As the United States national park system expanded, and the National Park Service was created in 1916 to oversee its management, the Forest Service reacted to its competitor in the land management arena by placing new emphasis on public recreation (see Hendee *et al.*, 1990, p. 35). Eventually this competition, in combination with the concerns of a few preservation-minded Forest Service employees – notably Aldo Leopold – led to administrative designation of Wilderness areas within National Forests. The first such Wilderness was the Gila (232 295 ha in southwestern New Mexico), which was established in 1924 (Nash, 1978, pp. 34–35; Hendee *et al.*, 1990, pp. 34–36).

Perhaps it was the combination of competition with the National Park Service, a Forest Service emphasis on conservation and 'scientific management' of natural resources, and increasing awareness of ecology (see Nash, 1978, pp. 195–6) and of special resources needing protection that led the Forest Service to begin administrative designations of 'research natural areas' (RNAs) in the 1920s. Development of the RNA system began in 1927 (Federal Committee on Ecological Reserves, 1977), with the designation of Santa Catalina RNA in Arizona (USDA Forest Service, 1972, p. 28).

Non-experimental research areas, such as RNAs, are much less likely to be altered or manipulated than the experimental type. They are 'reserved for scientific study of unusual or typical plant communities, soil or geological formations, or animal habitats' (USDA Forest Service, 1972, p. 28). RNAs are administratively designated by a number of federal agencies including the United States Forest Service, National Park Service, Fish and Wildlife Service and Bureau of Land Management. Other agencies, such as the Department of Energy and the Department of Defense, participate to a lesser extent. The system has been growing rapidly since the early 1970s. The Forest Service reported 94 RNAs in 1971, with a total of 42 855 ha (USDA Forest Service, 1972, p. 28). By comparison, Washington and Oregon alone now have about 100 Research Natural Areas administered by six agencies (see Franklin *et al.*, 1972; Greene *et al.*, 1986; Agee and Wason, 1987); more than half these areas are administered by the Forest Service. This designation of research land should continue to expand. In 1991, for example, New Mexico had 13 new Forest Service RNAs pending approval at the federal level (Dunmire, 1991, personal communication), and Shawnee National Forest in Illinois had nine areas designated and one pending (Illinois Society of American Foresters, 1991, p. 5).

Growth of the RNA system is probably rooted in the increased environmental awareness of the 1960s and early 1970s, and in the recently renewed expansion of environmental concerns. Not only did nature preservation become a more widespread and popular interest during these periods, but the realization of human impacts on the environment led to an interest in maintaining some relatively pristine areas for the purposes of comparison with more altered areas and monitoring environmental change. Scientists began promoting the view that certain areas should be set aside for research use:

> The value of wilderness to science, put baldly – very baldly, and not at all sentimentally – is the provision of study areas of pristine conditions.
>
> (Darling, 1960, p. 95)

> A civilized nation could, and should in my opinion, afford the maintenance of certain control areas for the study of natural variations in physical phenomena . . .
>
> (Leopold, 1960, p. 32)

When the Wilderness Act of 1964 (Public Law 88–577) legally formalized the wilderness system, which had existed solely as a result of administrative decisions within the Forest Service, it recognized several uses of Wilderness areas beyond the traditional purposes of recreation and conservation, including scientific use. Likewise, the Cascade Head National Scenic-Research Area in Oregon was created by federal legislation in 1974 (Public

Law 93–535) to provide for enjoyment of the area and 'to insure protection and encourage study of the area for research and scientific purposes' (USDA Forest Service, 1980). Mather (1991) has referred to the development of the 'post-industrial forest' in Britain and other developed countries where 'service functions such as recreation and environmental conservation are acknowledged alongside or in the place of wood production'. The recognition of additional service functions, indeed, extends to other than forested habitats, and environmental conservation itself encompasses a number of natural functions and the use of places for the expansion of knowledge.

A relatively recent programme, developed in the latter 1970s (Brenneman and Blinn, 1987, p. 1), involved the recognition of a need for long-term environmental research and the designation of specific sites for such research. Long Term Ecological Research sites, or Long Term Ecological Reserves (LTERs), are designated and funded by the National Science Foundation. At a time when researchers are faced with a professional atmosphere that tends to promote frequent contributions and thus short-term projects, these areas are exceptional in providing an arena where projects may last for many years. The LTER network of sites also supports research at a variety of spatial scales, including the collection of data on a regional or continental basis that can contribute to global change investigations (Swanson and Sparks, 1990). The system has continued to grow, and includes 18 sites owned and operated by a variety of agencies (see Franklin *et al.*, 1990 for an overview of the LTER programme).

To this point, research areas designated at the national level have been emphasized. Areas have also been recognized and designated for research use at lower levels of government, internationally and by non-governmental organizations. Among these are areas owned and controlled by universities, by preservation organizations (for example, the Nature Conservancy), by museums and similar institutions (for example, the American Museum of Natural History), and by professional organizations (for example, the Society of American Foresters). The major research areas mostly emphasize observational (non-manipulative) research, although experimental research is certainly pursued as well.

In the United States, a number of states have developed a system of reserved lands with explicit recognition of science or research as a purpose of the natural areas. For example, Washington State established a Natural Area Preserves System for research and educational uses (Nature Conservancy, 1977, p. 603). Wisconsin designated a number of 'scientific areas' on state (Department of Natural Resources) lands and on privately owned lands and areas controlled by other governmental bodies (Nature Conservancy, 1977, pp. 628–9). The Oregon State Land Use Act (1973) requires protection of 'ecologically and scientifically significant natural areas', among others (Lapping *et al.*, 1989, pp. 234–6). In California, a Natural

Reserve System was created in 1965 by the University of California group of universities. The reserves are for field research and teaching (see Kennedy, 1980; Kennedy, 1984; Gustafson, 1985; Ford and Norris, 1988). There are 26 reserves and four affiliated field sites in the California system, with a total of 52 610 ha. The University of California owns about 20% of this; the rest is included through use and management agreements with other organizations and individuals (Gustafson, 1989, personal communication). A combination of desires to preserve particular natural areas, and to provide for their use in education and science, is responsible for the development of most of these state natural/scientific reserve systems.

International recognition/designation of areas for research has been promoted by the United Nations' *Man and the Biosphere* programme. Areas that are viewed as important sites for ecological research are nominated by individual countries and designated as Biosphere Reserves by the United Nations. In support of preservation activities in less developed as well as more developed countries, Biosphere Reserves have three main purposes: research, conservation and education/training. The designation of a Biosphere Reserve often is in addition to other designations, such as National Park (for instance, Great Smoky Mountains National Park in the United States also is an international Biosphere Reserve). In 1988 the global network of Biosphere Reserves totalled 269 reserves in 70 countries, with nearly 1.43 million km^2 (see IUCN, 1985). There were 41 Biosphere Reserves listed for the United States, 13 for the United Kingdom and two for Canada in the *1985 United Nations List of Biosphere Reserves* (IUCN, 1985, pp. 32–9); most of the Reserves were in more developed countries and centrally planned economies (at that time), but a number were listed for Third World countries. In the United States, ecologically similar reserves are often 'matched': research in one area may emphasize observation and monitoring without manipulation (non-experimental research), whereas research in the other area may include manipulation and experimentation (experimental research) (see Franklin, 1979). For example, Coweeta Hydrologic Laboratory in North Carolina is the experimental counterpart of Great Smoky Mountains National Park, and the Jornada Experimental Range in southern New Mexico is the Chihuahuan Desert experimental counterpart of Big Bend National Park in Texas.

Discussion and Conclusions

Although many scientists, particularly ecologists, botanists and zoologists, see rural lands as a necessity for their research, and many preservationists try to publicize possible scientific or research values of areas in order to support preservation, the lands set aside for such activities are a little-recognized phenomenon of modern times. Analyses of the impacts of the designation of lands for research in rural areas, the extent of use of any such areas, and associated economic effects are lacking. In only a few cases can it be stated that a certain level of research use or research funding is associated with a particular area or system of areas. It has been reported that various organizations funded research in the Mount St. Helens area to a value in excess of US$6 million annually between 1980 and 1983 (USDA Forest Service, 1985, p. 95). In 1975 the National Park Service spent about US$5.5 million on its research programmes, but estimated total spending on research in the National Park System of at least US$16.5 million (National Park Service, 1976). Although these are not strictly research–designated areas, this gives some indication as to the use of certain types of protected rural lands for research: research areas that receive in excess of US$1 million annually include: H.J. Andrews Experimental Forest, Oregon (US$1.6 million), Coweeta Hydrologic Laboratory, North Carolina (US$2.2 million), and Hubbard Brook Experimental Forest, New Hampshire (US$1.2 million) (USDA Forest Service, 1985, p. 95).

Research dollars are highly important to the economies of western states in the United States, where huge areas in states like New Mexico, Utah and Idaho are designated and used for research purposes. With a conservative accounting of sites that can be clearly identified with research purposes and activities, about 10 425 km^2 of land used for research can be identified in the state of New Mexico. This 3.3% of the state's territory far exceeds the amount of land in urban and suburban land use categories. Research lands may be found in every state; for some states, such as New Mexico, Nevada, Maryland, Utah, Washington and Idaho, research lands (including experimental and test facilities) are significant in their geographic extent.

Some research lands are used for basic research in the biological sciences, geology and other scientific disciplines. Other lands are much more focused toward technology development. Many research pursuits, however, require rural lands. There are a number of reasons for this requirement:

1. The phenomenon being studied may exist only in places where there are relatively few people.
2. Developing an understanding of natural environmental conditions requires monitoring and baseline studies in fairly pristine areas. These studies allow comparisons with areas where human activities are certainly being felt and measurement of the extent of human impacts on the environment.
3. Some of the activities associated with development and testing of new technologies are considered unacceptable in areas with higher population concentrations.

Although land areas designated for research or scientific purposes date back 100 years or more (with the Hatch Act and individual university-related sites), they are mainly a rural phenomenon of this century. Reasons for designating research land have shifted as national and international concerns have changed. In the case of the United States, the sequence of emphasis in the designation of lands has been:

1. practical applications: agricultural-economic advancement;
2. preservation and basic science support: preserve natural areas of scientific interest;
3. practical applications: military-energy technology development;
4. preservation and basic science support: expansion of natural areas for research, monitoring and global change studies.

With an outlook of increasing populations, resource demand and pollution problems, perhaps the next step in the sequence will again be an emphasis on practical applications, as society attempts to manage the condition of the planet.

Designations, and the amount of land area devoted wholly or partially to research purposes, have continued to increase through the last few decades. There are numerous land reservation and designation systems that specifically set aside lands for research. In addition, multiple-use lands have also been designated for a variety of purposes, including research. Although there are physical and societal limits on the land available for research use, the designations of particular areas for use emphasizing research can be expected to continue for some time. These areas are likely to be protected in some way prior to designation for research. Within the United States, these additions are likely to be relatively small compared with some of the research areas that already exist. Through the years ahead, with the areas that already have been designated for some type of research, and the expansion of some of these areas, and the creation of new single research sites, research lands will continue to make up a significant proportion of rural land areas. They will be increasingly deserving of research attention themselves as a recent and expanding rural phenomenon.

Acknowledgement

The comments and encouragement of John A. Harrington, and the input of the editors and other conference participants, are greatly appreciated.

References

Agee, J.K. and Wason, C.R. (1987) *Stetattle Creek Research Natural Area*. USDA Forest Service Pacific Northwest Research Station, Portland, Oregon.

Bildstein, K.L. and Brisbin, I.L. (1990) Lands for long-term research in conservation biology. *Conservation Biology* 4, 301–8.

Brenneman, J. and Blinn, T. (1987) *Long-Term Ecological Research in the United States: A Network of Research Sites*, 4th edn. Long-Term Ecological Research Network, Oregon State University, Corvallis.

Darling, F.F. (1960) Wilderness, science, human ecology. In: Brower, D. (ed.), *The Meaning of Wilderness to Science*. Sierra Club, San Francisco, pp. 95–103.

Federal Committee on Ecological Reserves (1977) *A Directory of Research Natural Areas on Federal Lands of the United States of America*. USDA Forest Service, Washington DC.

Ford, L.D. and Norris, K.S. (1988) The University of California Natural Reserve System: progress and prospects. *BioScience* 38, 463–70.

Franklin, J.F. (1979) The conceptual basis for selection of US Biosphere Reserves and features of established areas. In: Franklin, J.F. and Krugman, S.L. (eds), *Selection, Management and Utilization of Biosphere Reserves*. USDA Forest Service Pacific Northwest Forest and Range Experiment Station, Portland, Oregon, pp. 3–27.

Franklin, J.F., Hall, F.C., Dyrness, C.T. and Maser, C. (1972) *Federal Research Natural Areas in Oregon and Washington: A Guidebook for Scientists and Educators*. USDA Forest Service Pacific Northwest Forest and Range Experiment Station, Portland, Oregon.

Franklin, J.F., Bledsoe, C.S. and Callahan, J.T. (1990) Contributions of the Long-Term Ecological Research program. *BioScience* 40, 509–23.

Greene, S.E., Blinn, T. and Franklin, J.F. (1986) *Research Natural Areas in Oregon and Washington: Past and Current Research and Related Literature*. USDA Forest Service, Pacific Northwest Forest and Range Experiment Station, Portland, Oregon.

Gustafson, S.S. (1985) *Natural Reserve System: The First Twenty Years*. University of California Natural Reserve System, Berkeley.

Hendee, J.C., Stankey, G.H. and Lucas, R.C. (1990) *Wilderness Management*. 2nd edn. North American Press, Golden, CO.

Illinois Society of American Foresters (1991) *Chapter Newsletter*. Illinois Society of American Foresters.

International Union for Conservation of Nature and Natural resources (IUCN) (1985) *1985 United Nations List of National Parks and Protected Areas*. IUCN, Cambridge.

Kennedy, J.A. (ed.) (1980) *University of California Natural Land and Water Reserves*. University of California Natural Land and Water Reserves System, Berkeley.

Kennedy, J.A. (1984) Protected areas for teaching and research: the University of California Experience. In: McNeely, J.A. and Miller, K.R. (eds), *National Parks, Conservation, and Development: The Role of Protected Areas in Sustaining Society*. Smithsonian Institution Press, Washington DC, pp. 538–45.

Lapping, M.B., Daniels, T.L. and Keller, J.W. (1989) *Rural Planning and Development in the United States*. Guilford Press, New York.

Leopold, L.B. (1960) Ecological systems and the water resource. In: Brower, D. (ed.), *The Meaning of Wilderness to Science*. Sierra Club, San Francisco, pp. 32–9.

Mather, A.S. (1991) Pressures on British forest policy: prologue to the post-industrial forest? *Area* 23, 245–53.

Nash, R. (1978) *Wilderness and the American Mind*. Yale University Press, New Haven.

National Park Service (1976) *Annual Report of the Chief Scientist*. US National Park Service, Washington DC.

Nature Conservancy (1977) *Preserving Our Natural Heritage: Volume II, State Activities*. Prepared for USDI National Park Service, Office of the Chief Scientist, Washington DC.

Schlesinger, A.M. (ed.) (1983) *The Almanac of American History*. Bramhall House, New York.

Swanson, F.J. and Sparks, R.E. (1990) Long-term ecological research and the invisible place. *BioScience* 40, 502–8.

USDA Forest Service (1972) *Report of the Chief 1970–71*. USDA Forest Service, Washington DC.

USDA Forest Service (1980) *National Forest System Areas as of September 30, 1980*. USDA Forest Service, Washington DC.

USDA Forest Service (1985) *Mount St Helens National Volcanic Monument: Final Environmental Impact Statement Comprehensive Management Plan*. Gifford Pinchot National Forest, USDA Forest Service, Pacific Northwest Region, Portland, Oregon.

15

Policy Options for the British Countryside

Andrew Gilg

The British countryside faces a time of great change in the 1990s, at both the macro and the micro levels. At the macro level the Uruguay round of GATT negotiations is attempting to achieve a massive dismantling of farm support policies throughout the industrial nations. Within Europe the European Community is however loath to accede to the very severe cut of 70% in the farm budget demanded by the Cairns group of agricultural exporting nations within GATT because they believe that this would bankrupt millions of small farmers. Accordingly they have offered a classic compromise, a 35% cut, with special dispensations for small farmers. Whatever happens it seems certain that farmers will face a decade during which prices will fall to unprecedented lows in real terms. This combined with continuing increases in productivity will mean that between 10 and 25% of all farmland will no longer be needed for agriculture, depending on which of the six business development strategies outlined by Bowler in Chapter 17 are followed by farmers. These changes will place severe strains on the rural economy and also pose deep questions about what to do with all the surplus land released from agriculture.

At the micro level, the continued restructuring of advanced economies, with industrial and service work increasingly decentralizing, will pose increasing demands for the development of rural land. As the need to preserve rural land for agricultural use recedes these pressures can only increase. Conservationists, however, are already learning to develop new arguments for preserving rural areas, for example, energy conservation and the need to cut rather than increase car mileage in order to reduce greenhouse gas emissions.

The Nature of Policy Making

Who makes policy?

There are many different theories and explanations for the way in which political power is held and implemented. For example, Abercrombie *et al.* (1988) have outlined four models of power and politics. First, there is the pluralist model which argues that power is dispersed and fragmented. Because no particular group is able to control the state or influence policy over more than a limited area, the role of the state is to act as a neutral referee. Second, there is the elitist model which argues that a number of key institutions and people dominate politics by only allowing certain issues to be debated. Third, there is the ruling class model which argues that capitalist employers form the most powerful elite of all, and that the only debate is how to run the economy as efficiently as possible. A fourth model, provided by a hybrid of the pluralist and elite models, is the corporatist model under which employers and trade unions form the two largest power blocs.

A rural perspective has been provided by Cloke (1987) who recognizes the pluralist and elite models, but also identifies two other models. First, the managerialist model in which professional agents of the state act as gatekeepers to policy making. Second, the structuralist, political economy, or Marxist model which suggests that class distributions represent the only real explanation of how power is exercised and policy created. Cloke has concluded that rural planning, far from being a rational process whereby clearly conceived objectives are enacted via channels of implementation, is in fact a search for consensus within prevailing arenas of power. As such, rural planning can make short-term incremental advances by adopting pragmatic measures for isolated issues, but the fundamental class injustices of rural areas remain.

Cloke and Little (1990), in a modification of this view, while still rejecting the pluralist model, do instead argue for an overlapping view of policy making based on aspects of the structuralist, elitist and managerialist models. This concurs with the view of Held (1984) who has concluded that political order is the outcome of a complex web of interdependencies between political, economic and social institutions and activities which divide power centres and create multiple pressures to comply.

This view seems to accord more closely with the real world than most of the theories outlined above, which suffer from the defect of all conspiracy theories. First, that decision makers are aware of the significance of current events, a very dangerous assumption, and second that any one group can actually predetermine the course of history. Four millennia of recorded history, in contrast, demonstrate that history is a random sequence of events, albeit shaped by formidable personalities, powerful ideas (Zeitgeiste), and underlain by the desire to amass capital. It is thus the standpoint of this chapter that policy formulation in the 1990s will be incremental, and accidental, rather than rational. Who, for example, in the summer of 1989 saw the collapse of Communism in Eastern Europe or, in July 1990, the Gulf War?

How is policy made?

Planning policies axiomatically imply rationality. In a utopian planning world, priorities would be listed and alternative ways of achieving the objectives identified by these priorities would be evaluated. The programmes and policies that best achieved these objectives would then be put in place. Unfortunately, in reality there are severe constraints on producing a rational plan for any activity, let alone something as complicated as the countryside. In general these constraints are our cognitive capacity to understand current events – stemming from both a lack of time and information – and a reluctance to widen the policy making process to others outside the magic circle. Within the countryside in particular, two recent studies of rural policy making (Baldock *et al.*, 1990; Collins *et al.*, 1990) have found that incremental decision making was forced on people by four key factors: (i) policy makers are faced with blocks of inertia imposed by extant policies; (ii) policies can only be gradually altered in a democracy; (iii) policy development is confined by fixed policy-making structures; and (iv) general uncertainty about the future leads to a natural reluctance to act. The net result is that policies are often a trade off between various interest groups, notably within the CAP, and reflect responses to immediate crises rather than any long-term strategy.

What is planning policy for, and how can it be implemented?

It is possible to identify three major roles for rural planning. The first role is to manage change by reconciling conflicts, notably over the physical development or change of use of land. The second role is to oppose change by designating sites where change is discouraged by preventing new land uses, or by helping to perpetuate existing uses. The third role relates to managing decline. This has become the main role of agricultural planning since about the mid-1980s, when the imposition of milk quotas signalled the end of post-war expansion.

Turning to implementation, there are, according to Blunden *et al.* (1985) and Selman (1988), five broad methods available, ranging from compulsion at one end of the spectrum to mere exhortation at the other:

1. public ownership or management of land via long-term leases;
2. regulatory controls, mainly negative, for example, planning permission;

3. taxation and fiscal disincentives to discourage undesirable uses;
4. financial incentives and tax breaks to encourage production or desirable land uses;
5. exhortation by providing advice, or practical demonstrations, and by threatening any of the more powerful devices above.

Variations on these five broad techniques are enormous, and Hill *et al.* (1989), for example, have identified 178 rural planning programmes running in the UK alone. However, the largest share of resources in these programmes was taken by agriculture, which consumed more than 90% of the funds available in the 1980s, a decade that is now considered.

Evolution of Planning Policies in the 1980s and Early 1990s

Space only allows a brief review here, but a fuller account is available in Gilg (1991a). In essence change occurred in the following five areas: (i) international and national change; (ii) agriculture; (iii) forestry; (iv) land use planning and; (v) conservation.

International and national change

Internationally British policy making became more entwined with Europe, especially after 1 January 1993 was set as the target date for a mass of harmonization exercises. Reform of the CAP, however, continues to be an insoluble problem and a resolution of this issue remains the number one item for rural policy makers in the 1990s, with expenditure under the CAP expected to rise from £17 000 million in 1989 to £24 000 million in the early 1990s.

Within the UK the biggest change has been the privatization of many public sector agencies, notably the Water Companies in 1989. Ironically, the watchdog authorities set up to monitor the new companies were given more power. Elsewhere the Environmental Protection Act (1990), EC directives on Environmental Assessment and other legislation on methods of controlling pollution introduced many more controls over farming, notably, straw burning (from 1993) and discharges of effluents from farms (as from 1991). For the future, the 1991 Nitrates directive threatens seriously to curtail arable farming in Eastern England by cutting nitrogen inputs to virtually zero.

Agricultural policy

The pressure to reform the CAP has forced a whole host of incremental knee-jerk responses, but no one big overhaul. Six key themes can be discerned. First, the 1984 imposition of milk quotas means that the number one income earning activity is now controlled in a benevolent straight-jacket. Second, a freeze on support prices has effectively imposed a cut in real terms, with prices being related to the amount produced in any one year for most arable products. Third, a gradual erosion of production grants has taken place with a transition to grants for either diversification or for managing the landscape. Fourth, there has been an increase in socio-structural aid via more and better financial aids and via an extension of qualifying areas and activities. Fifth, an attempt has been made to take 20% of arable land out of production by a Setaside Policy (see Chapter 7). Sixth, a series of measures have been introduced to pay farmers to manage whole areas of land for conservation, most notably in the Environmentally Sensitive Areas, and the Countryside Stewardship Scheme.

Among all this chaos, the Agriculture Act (1986) made a laudable attempt to impose some order by giving Agricultural Ministers the duty to endeavour to achieve a reasonable balance between: (i) a stable and efficient agricultural industry; (ii) the economic and social interests of rural areas; (iii) the conservation and enhancement of the countryside; and (iv) the promotion of the enjoyment of the countryside by the public.

Forestry

Within forestry there has been a major shift from the public to the private sector and a gradual expansion of the planting programme (see Chapter 13). Several grant schemes came and went in an attempt to increase private planting, but these were often undermined by contradictory changes in the arrangements for taxing forestry assets. In the early 1990s, however, two clear policy strands have emerged. First, further expansion of commercial softwood plantations in the uplands of Scotland and, second, a new policy aimed at reafforesting lowland England with amenity broadleaves, funded by planting grants and income supports for farmers diversifying out of agriculture.

Land-use planning

In spite of many attacks on the philosophy of planning during the decade, its power and methods remained remarkably intact. In essence the debate centred around whether cities should be allowed to extend more freely into the countryside. Battle was joined in the mid–1980s when two draft Department of the Environment Circulars proposed releasing more land for housing and taking away the strong presumption against development in the Green Belt. The self-interest of Tory (Conservative Party) voters in the Shires convincingly won the day, and in spite of a few Pyrrhic victories over converting farm buildings and allowing the limited expansion of many villages, the development industry was well and truly seen off by the NIMBY (not in my back-yard) forces. Indeed the amount of Green Belt land doubled in the decade, and in 1992 a limited form of design control over farm buildings was introduced.

Although development control retained its form, plan-making came under a perpetual cloud of uncertainty. Structure Plans and County Councils came to be seen as increasingly irrelevant, and although they were given a reprieve in 1990, the major adventitious reforms that will follow from the collapse of Thatcher's Poll Tax, in the Spring of 1991, will surely demonstrate once again that it is the 'Cock-up' rather than the 'Conspiracy' view of history that is correct.

Conservation

The big event of the decade was the 1981 Wildlife and Countryside Act. Although this marked a big step forward in the protection of species, it introduced a very controversial element into site protection, namely the payment of compensation to land users denied permission to change their use. This principle has been the cornerstone of subsequent attempts by the Conservative Party to conserve the landscape and wildlife. In essence this is to exhort land users to be conservationists, and when this fails to buy them off with compensation. This is of course the complete opposite to the system practised by land-use planners under the Town and Country Planning Acts. Not surprisingly, therefore, this clash promises to be one of the most controversial policy issues of the 1990s.

Review of Policies Proposed for the 1990s

Policy proposals for the 1990s have been gathered together by Gilg (1991b) from an exhaustive search of 1500 documents produced by rural organizations, pressure groups, academics and individuals. Table 15.1 presents the most popular policy proposals classified according to the five broad types of approach available to rural planners outlined earlier. The rest of this section discusses the policy proposals made in Table 15.1 but only for those policies advocated by six or more proposers because of space considerations. The discussion is based on the author's interpretation of all the evidence scrutinized in the 1500 source documents.

Public ownership or management of land

This concept is markedly out of fashion, but an extreme reform of the CAP, which led to a collapse of agriculture and thus of land prices, would

Table 15.1. Policy proposals most often advocated.[a]

Public ownership or management of land:	
6	Reintroduce social ownership of land
5	Greater use of land assembly powers
3	Compulsory move to smallholdings
Regulatory controls:	
19	Introduce planning controls over forestry
18	Introduce planning controls over farm buildings and/or farming operations
9	Introduce further quotas in agriculture
9	Allow new settlements in the countryside
8	Increase range of uses allowed in Green Belts or reduce their extent
6	Introduce zoning into development control
5	Restrict farm aid to certain upper limits
5	Impose controls on agricultural inputs
5	Make any form of farm aid dependent on an approved farm plan
5	Reform the consultation process in forestry
5	Relax planning controls a little
5	Release a lot of land for growth
5	Introduce two-stage public inquiries
3	Relax planning controls on re-use of farm buildings
3	Grade all areas by suitability for forestry
3	Ban new forestry on poor or sensitive sites
3	Introduce a third party right of appeal into land-use planning
Taxation and fiscal disincentives to discourage production or unwanted uses:	
16	Continue to cut back price support for farming
11	Reform of Management Agreements and Compensation system under 1981 Act
9	Introduce Landscape Conservation Orders
5	Impose taxes on agricultural inputs; for example nitrogen
5	Abolish all forestry subsidies
4	Abolish all tax concessions for forestry
4	Remove price support from agriculture
3	Move to world prices for agricultural produce
Financial incentives and tax breaks to encourage production or desirable uses:	
17	Scrap farm grants for production and replace them with conservation grants
12	Further encourage less intensive/organic farming
8	More environmental input into Extensification
6	Transfer agricultural aid to farmers not farming
6	More help for farm woodland
4	More use of Premium Conservation Grants in agriculture
4	More structural aid to farmers and rural areas
4	Move to payments by results (e.g. number of wild flowers produced)
3	Increase size of LFAs and extent of aid within them
3	Provide grant aid for creating new countryside
3	Give tax concessions for desirable land uses

Table 15.1. *Continued*

Exhortation, advice and community action:	
13	Set up Regional Plans/Strategies
6	More use of land for recreation
6	More emphasis on multiple-purpose forestry
5	Set up a system of Countryside Audits
5	Further develop National Planning Guidelines
4	Improve existing arrangements for integrating planning with other uses
3	Create Community Forests run by local groups/parishes
Administrative arrangements:	
13	Set up Regional Planning Authorities
10	Create new Ministry/Committee for Rural Affairs/Countryside
7	Set up a different system of Designated Areas
5	Implement Common Land reforms
4	Create more Designated Areas
4	Review system of Designated Areas
4	Modify Forestry Commission
3	Merge Countryside Commission and Nature Conservancy Council
3	Modify Ministry of Agriculture
3	Combine Counties and/or Districts
3	Create an Environmental Agency
3	Reform democracy by devolution
3	Set up more Environmentally Sensitive Areas

[a] The number is the number of experts advocating a proposal; the cut-off point for this table was three proposers per proposal.

provide a golden opportunity for conservation groups, with or without government aid, to purchase amenity sites at knockdown prices.

Regulatory controls

Since forestry is bound to expand as the main alternative to farming, planning controls are now urgently needed to oversee a land-use change which already this century has been as great as that of urban development.

The introduction of controls over farming as it declines may now seem less urgent, but at least half the country will still be the potential subject of hi-tech farming and here the need will become more urgent. For the sake of consistency, planning controls should be imposed everywhere, leaving Statutory Instruments to define exactly what should be controlled and where.

Quotas, in contrast, may have had their day. They are very effective short-term measures for tightly controlled systems like milk production, but cannot work for arable crops in general. They should be phased out as price reductions take their toll.

As farmland falls idle, the pressure to build in the countryside will grow. This pressure can best be deflected by boldly allocating land for large new settlements linked together by energy efficient transport systems. At all costs the pressure to revert to the unplanned sprawl of the 1930s must be resisted.

Taxation and financial disincentives

Policies to cut back price support are not only widely supported but are also the subject of the current (July 1991) proposals by the EC to reform the CAP. Under these proposals cereal support prices were planned to be cut to a target price of £70 a tonne – a cut of around 35% – milk quotas were planned to be cut by 4%, milk prices by 10%, and beef intervention prices by 15%. Controversially, the proposals also planned to keep farming viable by actually increasing spending – albeit at a slower rate – but on farmers not production. Accordingly, a social component was also proposed by which small farmers would be compensated for lowered prices and output, whereas big farmers would have to set aside land in order to qualify for intervention. These proposals have been widely criticized for subsidizing small inefficient farmers, fossilizing out-of-date farm systems, and continuing the chaotic bureaucracy of the CAP. It is hard to see these proposals having any long-term credence. Only severe cuts in overall spending on the CAP will do, and its proportion of EC spending must fall from over 60% to the 10–15% justified by the population of rural areas. Only then can a long-term agenda be set for the countryside.

In a similar vein the money spent on Management Agreements, although tiny compared with farm spending, is only a stop-gap as well as being a deeply flawed principle. By the same token moves to introduce Landscape Conservation Orders, and all associated proposals, should also be resisted. All they do is maintain a discredited system based on subsidized agriculture.

Financial incentives and tax breaks

Scrapping farm grants for production and replacing them with conservation grants for all the countryside is a utopian but hopelessly optimistic idea. The current proposal to subsidize farmers rather than farming is itself a wildy unpopular and untenable option, given the need for resources to be diverted to all sorts of other desirable programmes. If we assume that rural spending is cut to the 10–15% level commensurate with rural populations, then some of this much-reduced total could be spent directly on conservation. Europe-wide, the CAP is set to spend around £24 000 million in the early 1990s, or around 75% of an expected EC budget of £32 000 million. If the CAP's share were cut to 15%, this would cut

CAP expenditure to around £5000 million, and if the UK were given its proportional share this would give UK expenditure of around £800 million. If half of this were given to structural and social aid, this would leave £400 million for conservation, 92 times the £4.3 million allocated to the Countryside Stewardship Scheme (Countryside Commission, 1991). This could fund conservation payments on 1.6 million ha – 10% of all rural land – assuming a payment rate of £250 per hectare. Alternatively, the £400 million could be spent on a 'conservation-by-results' scheme, using the techniques pioneered in the Peak District National Park where farmers are paid for the number of wild flowers or butterflies counted.

Less intensive/organic farming can, in the short term, be encouraged by transition grants, but for the long term the demand thought to exist should allow around 20% of land to be farmed organically. The thought of returning to a 'Trumpton' type existence of small farmer/food producers selling to speciality shops may seem romantic, but it is a system that still exists in large parts of Mediterranean Europe. With greater leisure and higher expectations from food, going backwards may in fact be going forwards in terms of quality of life for both producers and consumers.

Extensification, like many of the other ideas related to existing CAP policies, is deeply flawed both in principle, and also because it deals with the side-effects of oversupported agriculture. It is not a question of driving the old machinery more slowly, but of inventing a new machine. Extensification related to organic and sustainable agriculture is, however, a different matter and one that can be supported. However, as Tarrant and Cobb show in Chapter 11, extensification is not as yet widely supported by farmers except in marginal areas.

Transferring agricultural aid to farmers not farming is at the heart of the current debate, but it is a short-term welfare measure while new structures emerge in rural areas. It would be very unwise to continue it, both on the grounds of economic efficiency, and in terms of providing any form of raison d'être for rural people.

Farm woodland aid is an excellent idea for Britain with its poor tree cover. However, the process of reafforestation needs to be closely monitored via democratically accountable development control, and accompanied by public access to private land. Reafforestation will also need to overcome the high level of farmer resistance noted by Ilbery (Chapter 7).

Exhortation, advice and community action

Regional Plans are a logical reaction to the way we live now, but they would be of little use without democratically elected Regional Councils and the power to exercise development control over major developments.

The greater use of land for recreation and multi-purpose forestry are highly desirable aims, except perhaps for landowners who may need more than persuasion to change their ways.

Administrative arrangements

Regional Planning Authorities have been discussed above. The creation of a Ministry of Rural Affairs is an attractive idea, but one that would have to overcome sectoral jealousies within Whitehall, and the increasing interconnectivity between urban and rural areas.

The system of Designated Areas has grown up incrementally over the years and, as with Whitehall Ministries, self-interest will ensure survival of the present system, especially since the 1991 review of National Parks recommended no major overhaul. It is to the self-interest of politicians that attention is now finally turned.

Political Scenarios for the 1990s

There are five broad political scenarios for the decade, but of course they are not mutually exclusive and they may occur in any sequence. Taking them in turn from the political right to left, but bearing in mind that this Chapter was written before the British General Election of April 1992, they are as follows.

A rebirth of Thatcherite conservatism

'Thatcherism' was in fact relatively muted, considering that there were – and still are – a group of people calling for radical reforms and the cutting back of all forms of government control. If this group managed to regain the initiative from the 'Major' moderates and, more problematically managed to get elected, then there would eventually be a return to a free market in agriculture with a host of unforeseen and largely unforecastable effects. Most likely, however, would be a division of the countryside into areas of hi-tech farming and abandoned land where the pressures to release land for urban growth would not be resisted by the 'New Right'.

Classless conservatism

It is much more likely that muted Thatcherism – as remodelled by John Major – will continue, both within the Party and in the electorate. If this is the case then muddling through will continue, and farm support will gradually be replaced by support for alternative employment or land uses. Planning controls if anything will get a little tighter as the Green imperative impinges on the free market. As urban-based employment continues to decentralize, the countryside will gradually evolve into the most prosperous and successful part of the nation. Protected by planning from unpleasant uses, it will become even more desirable to relocate there and thus a vicious circle will develop exacerbating the urban–rural divide.

Central coalition

Moving to the centre, the Liberal Democrats with large numbers of votes cast in local by-elections can still cling to the hope of a coalition. Ironically a coalition could be the most radical option of all. This is because the Liberal Democrats will make it a condition of any coalition that Proportional Representation and a devolution of real power to the grass roots takes place. Axiomatically a move to such a system will lead to a drastic reform of present structures and be irreversible since minority governments – the norm so far – could no longer make major changes. The system of planning would thus rapidly evolve to the 'bottom-up' Swiss model (Gilg, 1985).

Pragmatic Labour

While a coalition remains a pleasant mirage for Liberals, the Labour Party – after all the turmoil and policy reviews of the 1980s – are desperate to please the electorate and return to power. Their current policies are thus not very radical. They would almost certainly introduce planning controls over agriculture and forestry, but would retain support for rural areas, notably the uplands. They would also introduce regional planning.

Loves labour lost

Finally, the real Labour Party is alive but not well. Its addiction to central controls, corporate powers and national ownership have been discredited by the collapse of Communism, and given away by Thatcherism. Nonetheless, if the 1990–91 recession were to continue, then centralized controls might once again seem appropriate as a crisis measure. In such a scenario, a 1947-style set of reforms would include the imposition of controls over most land uses, quotas and powers to assemble land for the purpose of development. The driving force behind these controls would be the longstanding chimera of a National Land Use Plan based on the rational exploitation of resources as identified by surveys.

Conclusion

In conclusion, two issues will dominate the 1990s. First, reform of the CAP will remain the major challenge. However, it is hard to see how a radical reform can be achieved given the self-interest of those involved in the reform process. In the longer term, the re-entry of Eastern Europe and the dismantled Soviet Union into European affairs may be the lumpen straw that finally breaks the back of not just the CAP but of Europe–12, although this now seems less likely, post Maastricht. Second, the countryside will become more like the town, and indeed sometime during the next century towns may well reach rural settlement densities. How to achieve this transition without undue stress in inner urban areas and overuse of energy presents the biggest ever challenge to planners, especially in the absence of any clear statement from government about the role of the countryside in the 1990s, notwithstanding the failed attempt in the 1990 White Paper 'This Common Inheritance' (O'Riordan, 1990).

References

Abercombie, N., Wade, A., Soothill, K., Urry, J. and Walby, S. (1988) *Contemporary British Society*. Basil Blackwell, Oxford.

Baldock, D., Cox, G., Lowe, P. and Winter, M. (1990) Environmentally Sensitive Areas: incrementalism or reform? *Journal of Rural Studies* 6, 143–62.

Blunden, J., Curry, N. and Turner, A. (eds) (1985) *The Changing Countryside*. Croom Helm and the Open University, London.

Cloke, P.J. (ed.) (1987) *Rural Planning: Policy into Action*. Harper and Row, London.

Cloke, P.J. and Little, J. (1990) *The Rural State: Limits to Planning in Rural Society*. Oxford University Press, Oxford.

Collins, N., Bradbury, I.K. and Charlesworth, A. (1990) Formulation of the European Community price review: models of change. *Journal of Rural Studies* 6, 163–73.

Countryside Commission (1991) *Countryside Stewardship*. Countryside Commission, Cheltenham.

Gilg, A. (1985) Land use planning in Switzerland. *Town Planning Review* 56, 315–38.

Gilg, A. (1991a) Rural planning policies in the 1980s: an aide-memoire. *Progress in Rural Policy and Planning* 1, 45–58.

Gilg, A. (1991b) *Countryside Planning Policies for the 1990s*. CAB International, Wallingford.

Held, D. (1984) Power and legitimacy in contemporary Britain. In: McLennan, G., Held, D. and Hall, S. (eds), *State and Society in Contemporary Britain*. Polity Press, Oxford, pp. 229–369.

Hill, B., Young, N. and Brookes, G. (1989) *Alternative Support Systems for Rural Areas*. Wye College, Ashford.

O'Riordan, T. (1990) One-and-one-half cheers for Chris Patten. *Ecos* 11, 2–7.

Selman, P. (1988) Rural land use planning – resolving the British paradox. *Journal of Rural Studies* 4, 277–94.

V

Sustainable Agriculture as a Policy Option

16

The Policy Agenda for Sustainable Agriculture

John Pierce

As much as the current agricultural debate centres on the decline in world grain markets and the unsustainable financial costs of agricultural protection, it should not overshadow a number of other equally important and pressing farm problems affecting the long-term sustainability of agriculture. For example, the growth in productivity and accompanying high levels of sufficiency has generally been at the expense of the achievement of a number of other important goals, such as adequate income, community stability and resource and environmental conservation. The current crisis in agriculture, therefore, reflects a malaise in the industry that is far deeper than the problems of overproduction and excess capacity.

The objective of this chapter is to redefine the farm problem as a problem of balancing the numerous and complex goals that are now advanced in the name of sustainable agricultural production. After an examination of the interpretations of sustainability, some of the major issues and trade-offs in the pursuit of high-input agriculture are evaluated. For OECD countries, factors supporting unsustainable practices and the lack of integration between agricultural and environmental policies are then traced out. The discussion concludes with a consideration of revised standards for sustainable agriculture and the policies and programmes necessary to achieve those standards.

The Semantics of Sustainability

A profusion of terms is now used to describe agricultural operations and systems that are more environmentally positive and conserving of natural resources than prevailing mono-cropping systems. It is common to see terms such as biodynamic, humus farming, organic, permaculture, regenerative, natural, alternative, biological and low-input farming systems. Implicit in these terms is the recognition that farm operations function as agroecosystems within much larger and more complex ecosystems (Gliessman, 1990). In searching for a set of production practices that are common to the agroecosystem approach, a number of features can be identified. For example, some typical cannons are: the use of crop rotations, nutrient recycling, integrated pest management, control of biological cycles from within, multiple cropping and crop diversification, integration of crop and livestock production, high energy efficiency and optimal strategies that emphasize maximizing profit not production. Most of these features are illustrative of specific production practices. An important question that needs to be raised therefore is: can these features, individually and collectively, assist with the attainment of the larger objectives that society sets for its agricultural sector, land use and rural environment in the name of sustainability?

The new-found importance attached to sustainability represents the convergence of different forces reflecting, on the one hand, society's recognition of rapid changes to the quality and quantity of natural and environmental resources and, on the other, the political necessity to act with respect to those changes. A consensus is emerging that no single objective function exists for the achievement of sustainable agricultural systems. Sustainability is a multidimensional concept containing many diverse elements and goals. Accordingly, composite definitions of the concept are to be preferred. Brklacich *et al.* (1990, p. 10) define a sustainable food-production system as an 'agri-food sector that over the long-term can simultaneously: (i) maintain or enhance environmental quality; (ii) provide adequate economic and social rewards to all individuals and firms in the production system; and (iii) produce a sufficient and accessible food supply'. To these goals of sustainability Bowler, in the following chapter, has suggested the addition of budgetary and political criteria: from a public policy perspective, sustainable agriculture must be financially supportable and acceptable to the polity. Hence the concept is expressed in ecological as well as sociopolitical terms although, as Gliessman (1990) emphasizes, it is the ecological dimension that acts as the basic building block upon which the other elements of sustainability rely.

In many OECD countries, sustainability is interpreted not as an umbrella concept or mega-goal but, instead, independently as an ecological principle, placed alongside many other objectives of agricultural policy, such as: (i) income maintenance; (ii) economic development and efficiency; and (iii) production of adequate food and fibre (OECD, 1989). Regardless of the level of comprehensiveness attached to the theory and practice of sustainability, the challenge facing society lies in reconciling the diverse goals of stewardship, sufficiency/adequacy and equity, in understanding the trade-offs involved and in creating the necessary and sufficient political and economic conditions for achieving these goals.

Sources of the Farm Problem

Concerns have been expressed about the economic, social and ecological viability of modern agriculture (Norgaard, 1984; Smit and Brklacich, 1989; MacRae *et al.*, 1990; Pierce, 1990). Governments have become actively involved in regulating input and output of agricultural production yet the economic health of many operations remains in doubt. In order to identify the principal factors responsible for these conditions, it is necessary to rework some of the material presented in Chapters 1 and 2.

Market factors

As agriculture has become 'industrialized' through its dependence on increasingly concentrated agribusiness operations, and its pursuit of economic efficiency. This has led to increases in farm size and capital intensification, increased specialization, dependence on non-renewable resources and the marginalization of smaller units (Troughton, 1991). Aiding and abetting these processes has been technological change which is scale biased (Meilke and Warley, 1990). Inevitably the agricultural landscape has become dominated by a relatively small number of producers. During the late 1980s, in Canada, 25% of farm operations produced 75% of food output. In the United States and EC, 17 and 25% of the farms produced 70% and 60% of food output respectively (Meilke and Warley, 1990).

While increasing concentration of capital and economic activities (both at the farm and agribusiness levels) can lead to inefficiencies in the use of resources, there are of course many other instances of market imperfections and failings. A voluminous literature exists on these failings of the market, where private and social costs diverge and where poor or non-existent markets for public goods and services leads to misuse of resources (Redclift, 1987; Barbier, 1989; Daly and Cobb, 1989). It is unnecessary to review these arguments other than to make a few selected observations that have special relevance to agriculture. First, farmers' choice of crop, technology and intensity of production contribute to a variety of environmental costs or externalities (see Chapter 9). Troughton (1991, p. 78) suggests that '. . . the focus of stress seems to lie in the inherent contradictions between the non-renewable approach to the management of what should be renewable rural resources'. Land degradation poses on-site costs for the farmer and, potentially, for future generations. Market transactions for food production generally will not reflect the entire costs involved in the use of land resources. Public interests in preserving the productivity of the land may not be protected because the market may underestimate the social value of and future demand for land (in part because of the belief in the continued growth in technological change) and/or the farmer may not respond to its social value. In the case of off-site externalities (such as increased sedimentation of water bodies), the lack of markets for these products and the costs of control represent disincentives to the adoption of more environmentally positive approaches. Supporting this interpretation, Crosson and Ostrov (1990, p. 34) argued that; 'If the environmental advantages of alternative agriculture are real, the market system that fundamentally drives American agriculture will undervalue alternative agriculture relative to the conventional system.'

Second, with respect to the problem of the inability of the market to allocate public goods, it has been noted that since property rights are not attached to diversity, farmers tend to underestimate its social value relative to the social value of additional production. Clearly one of the main challenges will be to reduce the gap between 'farmers' economic interests in commodity values and society's interest in environmental values' (Crosson, 1990, p. 6).

Policy factors

To correct a number of real and perceived deficiencies in the operation of the market and to achieve a number of broad social and economic goals, government policy, macroeconomic and agricultural, has increasingly influenced the range of production options and the choice of these options available to the farm community. By manipulating or influencing what Batie and Taylor (1989) termed 'opportunity sets', government policy has increasingly affected the choices available to farmers regarding type of crop, inputs, level of intensity and marketing strategies.

Since the late 1940s agricultural policy in OECD countries has taken aim at the need to increase production and productivity, while at the same time improving the stability and income of the farming population (OECD, 1989). Together these far-ranging but patchwork policy measures have influenced the cost of inputs, the prices received for outputs and the profit margins of production (Table 16.1). Throughout Europe, Canada, the United States and Japan agricultural policies have encouraged increased specialization and intensification on the part of farmers, which in turn has led to numerous unintended and unanticipated consequences (Fig. 16.1). For example, the land resource base has been valued almost exclusively for its commodity function as a producer of food, with little recognition given to the provision of other important goods and services. Despite these issues, much of the concern today is still focused on the problems of the economics of overproduction and not the ecology of overproduction (Nijkamp and Soeteman, 1988).

The principal beneficiaries of these policies, at least in OECD countries, have been producers, in terms of lower operating costs or higher returns, protection from foreign competition and significant income transfers, and consumers in terms of lower costs and increased levels of self-

Table 16.1. Forms of government intervention within the agri-food sector.

Type	Explanation
Price/Income	
Safety net/deficiency payments Target prices Crop insurance	Ensures a minimum return to farmers based upon previously agreed prices for products and base acreages and yields.
Input	
Energy/chemical subsidies Transport subsidies Water subsidies Credit/tax Regional development	Influences the cost of inputs, their relative allocation and the competitiveness of operations. Generally leads to increased intensification and a narrowing of the range of crops/livestock produced.
Market	
Supply management Quality control Tariffs	Influences the quantity, quality and price of food marketed. Net effect of these approaches is to raise the domestic or threshold price to exceed the border or international price.
Research	
Production Resource	Contributes to a lowering of costs of production and increasing returns per unit of input. At the same time it limits choice of production technology and resource strategies.

sufficiency. The costs in terms of income transfers to farmers have been in billions of dollars – US$24 billion in the US, C$3.4 billion in Canada and 14 billion ECU in the EC in 1986. Producer subsidy equivalents for the same year were 36, 43 and 50% of total receipts for the United States, Canada and the EC respectively (Blandford, 1990). These amounts, which measure all transfers to domestic producers, are considerably higher than in the previous decade – often by a factor of two.

The economic effects and environmental implications of these policies are clear. Higher marginal costs of production associated with present levels of output can only be sustained by artificially high prices or subsidies on inputs. In numerous instances, commodity support programmes raise the price of crops above market clearing levels. As Girt (1990) has suggested, the consumer is supporting, through taxes in many instances, land management practices that accelerate soil erosion and other forms of environmental degradation. The behaviour of the farmer with respect to these policies is of course completely rational in terms of the short-term interests of the farming community. By ensuring higher returns to producers, either through lower costs or higher prices (or both), the value of

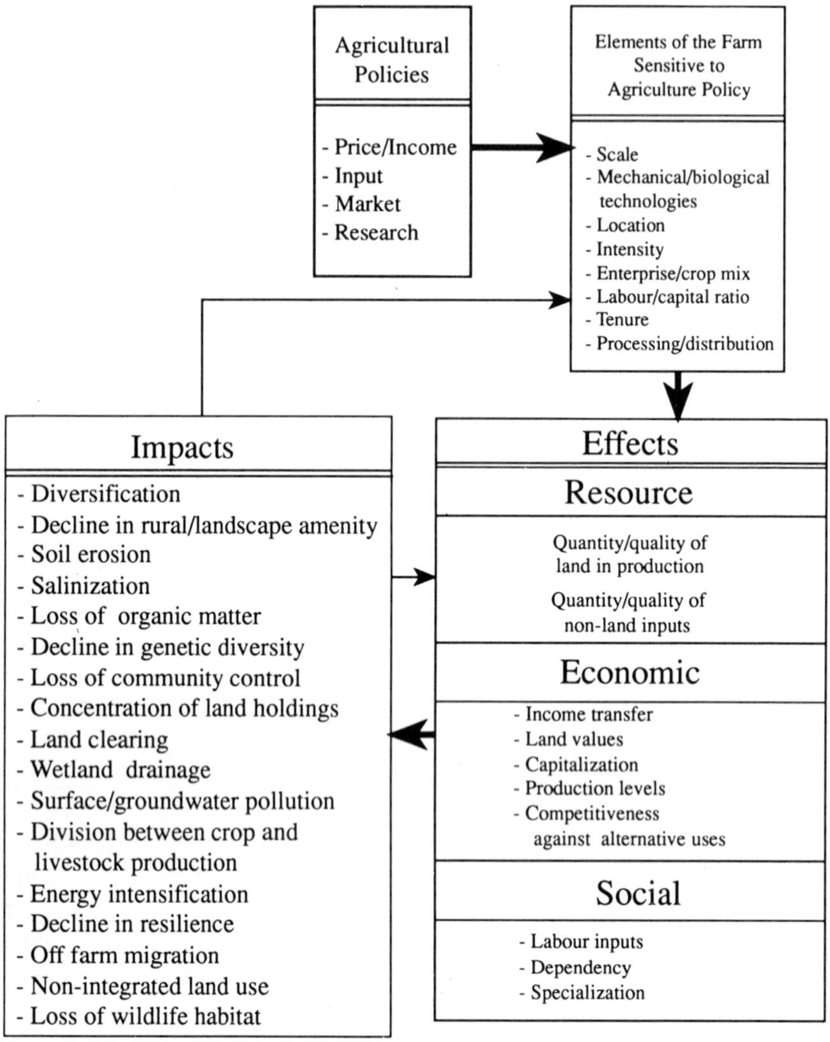

Fig. 16.1. Conceptualization of agricultural policy induced change to social, economic and natural systems.

agricultural goods and land is overstated. As a result, these conditions increase the opportunity cost of land remaining as wetlands or in alternative uses (Phipps and Crosson, 1986). The cost of removing land from production for conservation purposes is thereby increased by these measures. The social cost of policies to restrict pesticide and fertilizer use are also exaggerated (Lichtenberg and Zilberman, 1986).

Earlier it was suggested that specific policies tend to prescribe certain production options. In the process, potentially viable and less environmentally damaging approaches, be these low input, organic, or simply more diversified farm operations, have difficulty competing. Studies for Canada (Girt, 1990), the US (National Research Council, 1989; Faeth *et al.*, 1991) and the EC (OECD, 1989; Robinson, 1991) have identified a number of existing policies, programmes and standards that inhibit the spread and adoption of more environmentally benign agricultural practices and encourage farmers to ignore resource costs. These include research and extension programmes, base area requirements and cross-compliance provisions in crop commodity programmes, marketing boards and their marketing outlets, crop insurance, grading of food, health standards, target prices and pesticide licensing. While National Governments are beginning to recognize the need for change in these programmes, largely because of escalating costs and external pressures to free trade, farmers are justifiably hesitant to trade away long-established individual benefits, including security of the family farm, and shares of the market in return for a major restructuring in agricultural production with its expectant higher prices and more efficient deregulated markets. At the same time, few are willing to adopt alternative strategies and production systems that, in the absence of safeguards, may further weaken their economic stability. And it is recognized that the current economic situation cannot be improved over the long term without multilateral change – a change in the international rules that apply to all major players or stakeholders. The situation of course is not without promising developments. Although different approaches for decoupling have been proposed, the principle remains the same, i.e. that the income objective in agricultural policy should be separated from the production objective (Economic Council of Canada, 1988). By lowering or decoupling support to farmers, it is argued that the level of capitalization and intensification will decline, which in turn will encourage diversification and other risk-aversion strategies (OECD, 1989). It would also encourage the preservation of other values derived from the land resource base by reducing the disparity between the benefits of alternative uses. With the exception of New Zealand and Australia, no other OECD countries have proceeded with this strategy.

A number of very progressive steps have, however, been made to more closely integrate environmental and agricultural policy and thereby address one of the major issues in sustainable agri-food systems. As shown in Chapter 9 for the United States, the Food Security Act of 1985 contains sodbuster and swampbuster provisions preventing the extension of benefits to farmers who plough under highly erodible lands or drain wetlands for crop production. At the same time the Act authorizes, through the Conservation Reserve Program, the retirement of up to 18.3 million ha of highly erodible land and their placement in a 10-year conservation reserve. The 1990 Farm Bill has strengthened cross-compliance provisions and tied programme benefits to 1985 yield levels and reduced eligible base areas by 15%. The new Green Plan in Canada proposes a number of soil and water conservation initiatives. The Prairie Farm Rehabilitation Administration in Canada has instituted a Permanent Cover Program for certain grasslands and, related to this, the federal government, in concert with the provinces, has embarked upon a National Soil Conservation Program. In addition, Canada and the US are about to embark upon a C$1 billion 15-year Waterfowl Management Plan. In the EC the promotion of 'extensification' through setaside and financial compensation initiatives promises to reduce output.

Not to be ignored are non-profit foundations and organizations which are playing important roles in fostering better understanding of the environmental effects of agriculture, such as the US Nature Conservancy, the British Countryside Commission and the Canadian Soil Conservation Council. These groups are enjoying greater support from the public which, at the same time, is becoming increasingly conscious of the politics and ecology of food production (see Chapter 10).

These initiatives, nevertheless, must be placed in a wider context. For example, an analysis of the 1991 Canadian federal budget concluded that while C$296 million went towards the encouragement of environmentally sound farming practices, another C$1.5 billion went to the encouragement of destructive ones (Ralston-Baxter, 1991). The British experience is similar. Robinson (1991) determined that only 3% of the total Ministry of Agriculture, Fisheries and Food budget of £1200 million went to encourage environmentally friendly farming. For the EC as a whole, budgetary allocations in 1989 for market support were 28 million ECU compared with 1.4 million ECU for rural restructuring (Harvey, 1990).

Policy and Programme Possibilities

To achieve a better balance among the numerous objectives of sustainable agriculture, and to encourage the development and preservation of environmental services, will require changes in many of the now accepted rules, standards and conventions influencing the production of food and fibre. A key factor shaping these changes will be new institutions, policies and programmes that address not only the long-recognized imperfections in the operation of the market itself, but also the numerous distortions in the market which agricultural policy has created. The urgency of this task is made all the more important by the fact that the push for trade liberalization, through the Canada–US Free Trade Agreement, the North America Free Trade Agreement and the Uruguay Round of GATT, is undoubtedly creating a vacuum of uncertainty for the agricultural sector.

Since the publication of the *World Conservation Strategy* (International

Union for the Conservation of Nature, 1980), followed by the Brundtland Report, *Our Common Future* (World Commission on Environment and Development, 1987), governments have responded with a variety of proposals and plans of action to integrate environmental and economic decision making more closely in the name of sustainable development. For example, in Canada the National Task Force on Environment and Economy, an interprovincial committee designed to foster environment–economy integration, recommended that each province develop conservation strategies that can be used as blueprints for sustainable economic development (Canadian Council of Resource and Environment Ministers (CCREM), 1989). Conservation strategies are based on a number of important principles and processes (Table 16.2). These are the very principles that have been shortchanged by the legacy of post-war agricultural policy. If these principles are to have any practical value, they must be accompanied by a set of processes that deal with the definition, implementation and operation of such a strategy with respect to agriculture (Table 16.2). A number of these processes are detailed below.

Table 16.2. Key components of conservation strategies.

Guiding Principles
Resources are managed for sustainable use
Where possible, encourage environmentally sound renewable resource development over exhaustible non-renewable development
Keep society's options open
Integrate as much information as possible in planning and managing resources
Focus on causes as well as symptoms, and mix care with prevention
Find common ground
Base improvements to the resource management systems on present institutions and processes
Recognize that fairness and equity are essential to lasting change
Processes
Establish goals for sustainable development in accordance with the objectives and principles
Determine what is required to achieve these goals while accommodating both conservation and development objectives
Establish priorities for the goals using the criteria of significance, urgency, irreversibility, cost and probability of success
Identify the changes required to meet the goals
Analyse present and planned mechanisms for dealing with the prescribed changes
Propose the most cost-effective, least disruptive, and most integrated methods for facilitating change
Derive a plan of action for implementing the agreed-upon methods
Provide a means for a continuing evaluation of these goals, of progress toward achieving them, and of cost-effectiveness of the solutions

Source: Based on CCREM (1989).

Goals

The need to clarify goals associated with the production of food and fibre, as well as with other services derived from the use of the land resource base, cannot be overestimated. Both goals, and the priorities attached to these goals, will be the product of consultation among government departments, non-governmental agencies, farmers, consumers and citizen groups or, in short, as many stake holders as possible. Since there is no single objective that can define sustainable agriculture, in part because of the variation in resource and cultural contexts of the activity, one can expect considerable regional and national variation in the importance attached to food self-sufficiency, rural amenity, protection of wildlife habitat, tolerable levels of soil erosion, energy subsidy and rural community (see for example Harvey and Whitby's (1988) discussion of the competing nature of production and conservation goals in rural environments). An important ingredient informing these choices will be research, education and sharing of information about current problems, the sources of these problems and alternative courses of action.

Targets

If goals provide the general direction for policy development, targets or objectives provide the destination. Assuming stewardship is a goal, how does society define an acceptable level of conservation of resources? Some have proposed constancy of natural capital as a necessary condition for sustainable development (Pearce, 1988; Costanza and Daly, 1990). To operationalize such a concept for agriculture, it would be necessary to define measures of adequate foodlands, soil productivity, genetic diversity, water quality and dependence on hydrocarbons as a source of energy. In addition to this, objectives should be specified with respect to other goals such as equity and sufficiency. The former could be measured in terms of income distribution and the latter in terms of the balance between economic and nutritional demand for food. Even more challenging would be the development of collective measures of social income and welfare that incorporate depreciation of natural and human capital in national income accounts (Repetto *et al.*, 1989).

Linkages

The third important ingredient is the improvement of links and integration among existing institutions and the development of new institutions and/or organizational structures to foster these links and integration. As one report on the matter recently stated,

> Integration requires policy makers to give full consideration to and to accept responsibility for the effects of their policies on the objectives of all other sectors.
>
> (OECD, 1989, p. 56)

The interdependence of agriculture and the environment requires a greater acknowledgement of these responsibilities and means to achieve greater harmony of interests. Particularly important is the establishment of agricultural policy in tandem with or within the context of integrated land use and resource plans. In turn a much broader and more comprehensive perspective will have to be adopted with respect to accounting for non-agricultural values in both rural and urban regions and developing strategies for stimulating alternative agriculture. To achieve this a much greater level of political will and leadership will have to be brought to bear on the process than has been realized to date (Skogstad, 1991).

These initiatives are not limited to the boundaries of sovereign nations but must be transnational and multilateral. For example, a common concern being expressed with respect to the impacts of trade liberalization on agriculture is that unless all trading nations adopt similar environmental standards some countries will be at a comparative disadvantage with respect to others. Unfortunately, the end result of this process may be the design of strategies which are oriented to serving the lowest instead of the highest common denominator.

Procedures

The development of procedures for anticipating economic, social and environmental consequences of different policies that directly and indirectly affect the short- and long-term viability of agriculture and other related resource sectors is required. What are the economic advantages and disadvantages of a shift in incentives that favour conservation tillage; how will the decoupling of income support affect the prospects for alternative agriculture; how much land is required under alternative production strategies; what are the costs and benefits of pesticide-based and pesticide-free agricultural strategies; how will a decline in research spending for conventional agricultural practices impact that sector?

The listing of these questions belies the significant, if not daunting task of recognizing the complex interactive properties of agricultural systems. As a simple example, the use of artificial fertilizers, which is often subsidized, is a yield-augmenting but land-saving strategy. Reduction in fertilizer use could promote the use of greater quantities of land, some of which may be more hazard prone. Given these issues, there is a clear need to ensure, first, that proposed solutions must be neutral with respect to the generation of additional problems, and second, that society fully explores a wide range of scenarios relating to the future of agriculture and rural environments. In this regard, Robinson (1989) has proposed the use of prescriptive or normative approaches such as backcasting. Under such a system, the concern is with 'how desirable futures can be attained'. Once a series of desirable futures is defined, the procedure is to work backwards to determine the policies and procedures for achieving the goals as well as the implications of easy approach.

To assist in determining the relative merits of conventional versus alternative forms of production, criteria will be required 'that account fully for on-farm and off-farm environmental costs, and that estimate the comparative return to various systems free of the distorting effects of baseline agricultural policies' (Faeth *et al.*, 1991 p. 24). As is often the case, however, economic evaluations of alternative policies are rife with methodological problems. Typically there is disagreement over the stakeholders, shadow prices, time horizons, discount rates, performance standards, and general assumptions about technological change. For these techniques to be useful at resolving conflict, the values and underlying assumptions being used must be made far more explicit (Reichelderfer, 1991).

Programmes and instruments

The transition to sustainable modes of production will be complex and so will be the programme and instrument responses (Harvey and Whitby, 1988). OECD countries have already experimented with a wide variety of mechanisms that will more closely integrate environment and economy. In a few cases these have been designed to place alternative agriculture on an equal footing with conventional forms of food production. In a review of European policies, Ervin and Tobey (1990) identified: (i) incentive-based policies which include direct charges or fees for fertilizer and pesticide use, subsidies to encourage environmentally friendly practices, management agreements and setasides; (ii) regulatory instruments such as controls on the quantities of fertilizer, pesticide and manure use; and (iii) public investments. Although it is beyond the scope of this chapter to provide an in-depth review of the effectiveness of these and other policy instruments, some of the more important features of the first two approaches can be outlined.

Voluntary and incentive-based approaches to reduce the environmentally harmful effects of agriculture are used extensively in OECD countries. Farm extension services, soil conservation officials and pesticide management personnel can play effective advisory roles. Incentives offered for improvement in the agricultural environment or natural capital of agriculture are an important way to bridge the gap between conservation investments and conservation returns (OECD, 1989). In Canada,

Girt (1990) noted that, under the proposed North American Waterfowl Management Plan, an annual payment of C$39 per hectare would be sufficient to convert relatively large tracts of marginal cropland (1.8 million ha) to permanent cover and wildlife habitat. Clearly if sustainable development contains an important intergenerational dimension, then society must provide some added incentives. Internalizing the benefits from the supply of environmental services is an important step in that direction.

Financial penalties are also necessary to achieve specific objectives. The polluter-pays principle is attractive in theory but difficult to enforce in practice given the non-point character of much agricultural pollution. However, in cases of point sources, the principle can be effective. Applying taxes to fertilizers, pesticides and fuels can eliminate much of the waste associated with these inputs and assist in internalizing a larger share of the total costs associated with agricultural production. The use of marginal cost pricing for irrigation water is also an effective means to reduce waste and some of the environmental problems associated with excess use of surface and groundwater sources (see Chapter 9).

Involuntary approaches that use regulation or 'command and control' are often less efficient, since they ignore the sources of the problem. Nevertheless, they are easier to implement and can represent sector wide (minimum) standards and cross-compliance controls, whether it is in terms of soil loss, land-use mixes, groundwater contaminants, pesticides use, water withdrawals or emissions generally. Worth emphasizing is the fact that economic acceptability of regulatory approaches is dependent on the implementation of comparable instruments in other jurisdictions.

What remains unclear is the degree to which these instruments will have to accompany and support the agricultural environment, as trade becomes freer and as farm support is decoupled from production. Such changes underline the necessity for an ongoing review and assessment of the progress toward sustainable agriculture.

Review

Implicit in the establishment of on-going reviews is the determination of the degree to which policies are having their desired effects in terms of a number of the previously outlined goals and criteria. Unforeseen complications and interaction effects are bound to arise. This becomes all the more important over the intermediate to long term when one considers possible changes in the forces affecting the demand for and supply of food, and ultimately the adequacy of the agri-food sector. Pre-eminent among these forces are changes in technology, population and income levels, climate, geopolitical situation and the values society attributes to the agricultural sector. A central criterion for judging the effectiveness of new policies should be the degree to which the agro-ecosystem retains sufficient margins of resilience to withstand economic and biophysical shocks.

Conclusions

Instead of a single-minded commitment to adding value, society must move on to an agenda devoted to balancing values in the agri-food sector and the rural environment. From a policy perspective the main task ahead is to redesign and implement programmes that will encourage new 'opportunity sets' for the economic and competitive viability of the farmer and the agricultural system, while at the same time meeting the nutritional needs of society and ensuring the protection of important environmental services and amenity values. As much attention will have to be paid to the political and decision making apparatus necessary to achieve change, both nationally and internationally, as to the actual tools and mechanisms themselves. That decision making will have to reflect complex local and regional variations in needs and biophysical conditions as well as in the knowledge and aspirations of the individual farmer. In future, policies will have to be far more sensitive to local conditions and constituents if they are to hold the promise of supporting a transition to sustainable food systems. That promise is also contingent on pushing for large-scale and comprehensive change to the broader economic, social and institutional environments within which agriculture operates.

Acknowledgement

The author would like to thank the Social Sciences and Humanities Research Council of Canada for its financial support of this research.

References

Barbier, E.B. (1989) *Economics, Natural Resource Scarcity and Development.* Earthscan, London.

Batie, S.S. and Taylor, D.B. (1989) Widespread adoption of non-conventional agriculture: profitability and impacts. *American Journal of Alternative Agriculture* 4, 128–34.

Blandford, D. (1990) The costs of agricultural protection and the difference free trade would make. In: Sanderson, F.H. (ed.), *Agricultural Protectionism in the Industrialized World.* Resources for the Future, Washington, pp. 398–432.

Brklacich, M., Bryant, C.R. and Smit, B. (1990) Review and appraisal of concept of sustainable food production systems. *Environmental Management* 15, 1–14.

Canadian Council of Resource and Environment Ministers (CCREM) (1989) *Conservation Strategies: A Compendium of Canadian Experiences*. CCREM, Ottawa.

Costanza, R. and Daly, H. (1990) *Natural Capital and Sustainable Development*. A paper prepared for the Canadian Environmental Assessment Research Council, Workshop on Natural Capital, Vancouver.

Crosson, P.R. (1990) Supplying the environmental values of agriculture. *Resources* 98, 4–8.

Crosson, P.R. and Ostrov, J.E. (1990) Sorting out the environmental benefits of alternative agriculture. *Journal of Soil and Water Conservation* 45, 34–41.

Daly, H.E. and Cobb, J.B. (1989) *For The Common Good*. Beacon Press, Boston.

Economic Council of Canada (1988) *Handling the Risks, a Report on the Prairie Grain Economy*. Supply and Services Canada, Ottawa.

Ervin, D.E. and Tobey, J.A. (1990) *European Agriculture and Environmental Policies: Sorting Through Incentives*. A paper prepared for the Conference 'Is Environmental Quality Good for Business' American Enterprise Institute, Washington DC.

Faeth, P., Repetto, R., Kroll, K., Qi, D. and Helmers, G. (1991) *Paying the Farm Bill: US Agricultural Policy and the Transition to Sustainable Agriculture*. World Resources Institute, Washington DC.

Girt, J. (1990) *Common Ground: Recommendations for Policy Reform to Integrate Wildlife Habitat*. Environmental and Agricultural Objectives on the Farm, Wildlife Habitat Canada, Ottawa.

Gliessman, S.R. (1990) Quantifying the agro-ecological component of sustainable agriculture: a goal. In: Gliessman, S.R. (ed.), *Agroecology*. Springer-Verlag, New York, pp. 366–70.

Harvey, D. (1990) *European Policies and Laws Affecting Sustainable Agriculture*. A paper presented at the Globe '90 Conference, Vancouver.

Harvey D. and Whitby, M. (1988) Issues and policy. In: Whitby, M. and Ollerenshaw, J. (eds), *Land Use and the European Environment*. Belhaven Press, London, pp. 143–86.

International Union for the Conservation of Nature (1980) *The World Conservation Strategy*. IUCN, Geneva.

Lichtenberg, E. and Zilberman, D. (1986) Problems of pesticide regulation. In: Phipps, T.T., Crosson, P.J. and Price, K.A. (eds), *Agriculture and the Environment: Annual Policy Review 1986*. Resources for the Future, Washington DC, pp. 127–38.

MacRae, R.J., Hill, S.B., Henning, J. and Bentley, A.J. (1990) Policies, programs and regulations to support the transition to sustainable agriculture in Canada. *American Journal of Alternative Agriculture* 5, 76–92.

Meilke, K.D. and Warley, T.K. (1990) Canada. In: Sanderson, F.H. (ed.), *Agricultural Protectionism in the Industrialized World*. Resources for the Future, Washington DC, pp. 112–80.

National Research Council (1989) *Alternative Agriculture*. NRC, Washington DC.

Nijkamp, P. and Soeteman, F. (1988) Land use, economy and ecology. *Futures* (Dec.), 621–34.

Norgaard, R.B. (1984) Co-evolutionary agricultural development. *Economic Development and Cultural Change* 32, 525–46.

OECD (1989) *Agricultural and Environmental Policies, Opportunities for Integration*. OECD, Paris.

Pearce, D. (1988) Economics, equity and sustainable development. *Futures* (Dec.), 598–605.

Phipps, T.T. and Crosson, P.R. (1986) Agriculture and the environment: an overview. In: Phipps, T.T., Crosson, P.R. and Price, K.A. (eds), *Agriculture and the Environment: Annual Policy Review 1986*. Resources for the Future, Washington DC, pp. 3–31.

Pierce, J.T. (1990) *The Food Resource*. Longman, London.

Ralston-Baxter, R. (1991) *Environmental Assessment of the Federal Budget for the Department of Agriculture*. Resource Futures International, Ottawa.

Redclift, M. (1987) *Sustainable Development: Exploring the Contradictions*. Methuen, London.

Reichelderfer, K.F. (1991) *Agriculture and Resource Sustainability: Can Economics Help?* Resources for the Future, Washington DC.

Repetto, R., Magrath, W., Wells, M., Beer, C. and Rossini, F. (1989) *Wasting Assets: Natural Resources in the National Income Accounts*. World Resources Institute, Washington DC.

Robinson, G.M. (1991) EC agricultural policy and the environment. *Land Use Policy* 8, 93–107.

Robinson, J.B. (1989) *Exploring A Sustainability Future For Canada: The Next Step In The Conserver Society Discussion*. Department of Resource and Environmental Studies, University of Waterloo, Waterloo, Ontario.

Skogstad, G. (1991) *Political Institutions and a Sustainable Agriculture*. Science Council of Canada, Ottawa.

Smit, B. and Brklacich, M. (1989) Sustainable development and the analysis of rural systems. *Journal of Rural Studies* 5, 405–14.

Troughton, M.J. (1991) Agriculture and rural resources. In: Mitchell, B. (ed.), *Resource Management and Development*. Oxford University Press, Toronto, pp. 53–84.

World Commission on Environment and Development (1987) *Our Common Future*. Oxford University Press, Oxford.

17

'Sustainable Agriculture' as an Alternative Path of Farm Business Development

Ian Bowler

This chapter is concerned with 'sustainable agriculture' interpreted as a type of farm business development. The discussion is placed in the contexts of the contemporary international farm crisis, the restructuring of agriculture and the European Community (EC).

Agricultural Restructuring and the International Farm Crisis

The 1980s have witnessed the development of an 'international farm crisis' which has provoked a relatively rapid restructuring in the use of agricultural resources within most developed economies (Goodman and Redclift, 1989). Previous chapters have shown how the farm crisis is characterized by the escalating costs of state-financed farm price support programmes, the over-supply of domestic markets, rising farm indebtedness and bankruptcies, and the increasingly unacceptable environmental damage caused by modern farming practices. A consensus of scientific opinion now interprets the 'crisis' as the logical outcome of the industrial model of agricultural development – a model first established in North America but now diffused into all developed economies.

The farm crisis has become 'international' in two respects. First, the characteristic features of the crisis are evident in most developed countries, not least because food systems have become increasingly internationalized and interdependent. Second, certain countries, notably those in the EC, have subsidized the export of food products in surplus production onto world markets. This policy has brought the new food exporters, such as the EC, into conflict with countries which have traditionally supplied world food trade. The resulting international trade disputes have been carried into the Uruguay Round of GATT where, at the time of writing, an impasse exists in negotiations: proposed reductions in the level of protection offered to farmers in the EC are judged insufficient by those countries disadvantaged in international trade.

In common with many other developed countries, agricultural policy in the EC was progressively modified during the 1980s so as to meet the farm crisis by introducing a greater degree of price responsiveness. Beginning in 1982 with the application of guarantee thresholds and co-responsibility levies to milk and other products in structural over-supply, control measures were extended by quotas on milk production in 1984 (Cox *et al.*, 1990), 'stabilizers' to limit the quantity of production eligible for full price support and, most recently in 1988, arable setaside for cereal production (Ilbery, 1990; Jones, 1991). At the same time, the EC began to develop measures to protect the rural environment from the worst excesses of modern agriculture (OECD, 1989). The implications of these programmes have been reviewed for the EC and the UK by Potter (1986) and Robinson (1991) respectively, but it is too soon to evaluate their actual impacts. Already clear, however, is the marked spatial variation in the response of the farming community to these measures.

Emerging Farm Business Development Strategies Under the Restructuring of Agriculture

The competitive and atomistic structure of the farm sector has ensured that a variety of paths of farm business development (also termed 'adjustment', 'survival' and 'restructuring') are taking place in response to the international farm crisis. There is no unitary pattern of agricultural development. Simplifying, continuing farm businesses are following one or a combination of these 'paths of development':

1. extension of the industrial model of farm business development based on traditional products and services;
2. redeployment of farm resources (including human capital) into new agricultural products or services;
3. redeployment of farm resources (including human capital) into new non-farm products or services;
4. redeployment of human capital into an off-farm occupation (OGA, Other Gainful Activity);
5. maintenance of traditional farm production and services with either reduced inputs and/or reduced income;
6. hobby or part-time (semi-retired) farming.

A similar typology of farm business restructuring (termed 'survival strategies') has been advanced by Whatmore *et al.* (1987), but in both typologies there can be detected an underlying three-dimensional division in farm business trends – in effect between: (i) the maintenance of a full-time, profitable food production basis to the farm business (i.e. paths 1 and 2); (ii) the diversification of the income base of the business by restructuring resources into non-farm enterprises (i.e. paths 3 and 4); and (iii) the marginalization of the farm as a profitable business (i.e. paths 5 and 6). In the longer term, individual farm businesses are likely to make a transition from strategies (i) to (ii) and to (iii); more difficulty is likely to be encountered in making the transition in the opposite direction, owing to the need for scale economies and/or large capital investments to re-enter profitable full-time farming or diversified enterprises. Farm businesses remaining in pathways 1 and 2, for example, will probably have to absorb the consequences of the forthcoming 'biotechnology' revolution (Goodman *et al.*, 1987). Projections for this revolution include genetically engineered crops and livestock, a farm sector producing generic food components for the food-processing industries (for example, starch, glucose, vegetable protein), livestock production mainly within buildings, and 'value-added' food processing at the farm level. Under the 'biotechnology' scenario, a relatively few but very large production units would be able to supply most food requirements.

By comparison, farm businesses in pathways 3 and 4 seem destined to obtain an increasing proportion, but not all, of their income from non-farm enterprises either on or off the farm itself, and from the production of environmental goods and services. Both features are already established in Western Europe: as previously described, 'environmentally friendly' farming practices are being subsidized by many countries, whereas programmes to promote the diversification of farm businesses are in place (for example, the Farm Diversification Grant Scheme (1988) in the UK – see Chapter 7). Diversification trends in the UK reveal a bias towards urban fringe and tourist regions, with an emphasis on farm accommodation, equine enterprises and direct farm sales (Bowler and Ilbery, 1992).

The extent to which each of the six paths of development is followed by farm businesses will be determined, in large part, by state farm policies. In the context of the EC, for example, much will depend on the extent to which the emerging biotechnology is allowed to penetrate agricultural production. Conflict on this issue between environmentalists, industrial capital, animal welfare groups, consumer groups, farming interests and the state is only just developing. However, 'food health' has already become a politically sensitive issue. Powerful economic interests support the new technology and its industrial model of agriculture, since it offers scope for a new cycle of agroindustrial accumulation (Commission of the European Communities, 1989). Food processors, for example, view the developments as an assurance of the continuation of the supply of their cheap raw materials. Competition between countries, regions and individual producers is likely to ensure the adoption of developments in biotechnology, not least because food processors are now organized as 'food trans-nationals' and seek out their raw materials on a global scale (Bijman *et al.*, 1986). The new biotechnology seems destined to favour farm businesses following paths of development 1 and 2, and large farm units at the expense of small. Indeed the 1991 proposals from the European Commission on the Common Agricultural Policy (CAP) of the EC (The 'MacSharry Proposals' of July 1991) appeared to anticipate this evolving structure. Larger farm businesses would compete in the international market with significantly reduced levels of price support; smaller farms would receive most of the benefits of continued state support through direct income payments, and price supports which would be regressive with increasing volumes of production from each farm unit. Such a policy would deepen the already evident polarization in the farm-size structure and link that structure to the six 'paths of development' already described. For example, it is difficult to see how smaller farm businesses can escape the pressure to follow paths 3 to 6 in the typology previously outlined.

'Sustainable Agriculture' as a Path of Farm Business Development

In the context of the EC, the present political response to the international farm crisis does not appear to offer a stable future for agriculture in either the short or medium term. Attention is being focused on short-term economic actions to resolve the farm crisis (The Economic Synthesis Model) or partial regulatory actions to meet urgent environmental problems (The Managed Scarcity Synthesis Model) (Schnaiberg, 1975, quoted by Troughton, 1991). Farm diversification exemplifies the first model, but the available evidence suggests that this development offers 'niche' market solutions to only a proportion of farm businesses. The introduction of 'environmentally friendly' farming practices falls under the second model, but these are essentially palliatives for they do not change the essential nature of the agroindustrial model that produces the damaging environmental impacts (Bird, 1989).

In North America, by contrast, and as shown in the previous chapter, attention has already turned to examining the potentialities of 'sustainable agriculture' as a medium- to long-term alternative to agroindustry under an Ecological Synthesis Model. Pierce (1990), Troughton (1991) and Brklacich *et al.* (1990) have reviewed and introduced the relevant literature to geographers, although in origin the ideas can be traced to the late 1960s and early 1970s when societal concern with the use of natural resources first emerged (Merril, 1976). Applying this literature to a West European context provokes two problems: on the one hand, defining the concept of 'sustainability', and on the other hand applying the concept to particular farming systems. On the issue of the definition of 'sustainability', the discussion by Pierce (1990) and Brklacich *et al.* (1990) is incomplete for a West European context. They demonstrate that no single objective function exists for the achievement of a sustainable agriculture, there being at least three criteria to be met. Summarizing, these are: (i) environmental – to maintain or enhance the agricultural resource base; (ii) socio-economic – to provide equitable economic rewards to individual farms and rural communities in the production sector; (iii) productionist – to produce a sufficient food supply. Keeney (1989), for example, summarizes these criteria in the following definition of 'sustainability': 'agricultural systems that are environmentally sound, profitable, and productive and that maintain the social fabric of the rural community'. To these goals of 'sustainability', however, should be added: (iv) budgetary – to absorb an acceptable proportion of state (public) expenditure and (v) political – to maintain the political support of society. In Western Europe these two goals are particularly important: there are few prospects of state assistance being totally withdrawn from agriculture, yet the budgetary costs of programmes such as the CAP will have to be brought under control in the medium term; political support for the farm sector, both among politicians and voters, is necessary to sustain the farm programmes both on income supports (see ii) and the environment (see i). At present, however, political support among the population of Western Europe for agroindustry appears to be ebbing, if the results of successive opinion polls are to be believed.

Turning to the application of the concept of 'sustainability' to farming systems, to date the debate has been conducted largely in macroconceptual or theoretical terms (for example Hill, 1985) or else in the broad context of agroecosystems (for example Jackson, 1989). In the latter case, for instance, checklists of farming practices compatible with a 'sustainable agriculture' are widely available, although some writers emphasize that 'sustainable agriculture' is a goal rather than a set of prescriptions for exactly how agricultural systems should operate (Allen and van Dusen, 1989). Such checklists include: diversified land use, the integration of crop and livestock farming, crop rotations, organic manures, nutrient recycling, low energy inputs, low inputs of agrochemicals and biological disease control. One outcome of the application of such a checklist approach is a wide range of terms to describe a 'sustainable agriculture'. These include, but are not limited to: low imput-output farming, permaculture, alternative agriculture, organic farming, regenerative farming, ecological farming and biodynamic farming. So as to focus attention on the implications of these terms for farming systems, it is helpful to define three ideal types along a continuum of the increased application of ecological principles to the farm business.

1. *Diversified farming* (see also Chapters 6 and 7): the introduction of a variety of crops (including timber) and/or livestock into the farm business, including crop rotations and 'environmentally friendly' farming practices (environmental goods). Other farming practices associated with agroindustry need not be affected.
2. *Low input-output farming*: a reduction of purchased inputs per hectare of farmland, leading to less intensive farming methods with lower levels of production. Farming practices associated with agroindustry can be continued at a lower level of intensity, compensated by 'environmentally friendly' farming practices.
3. *Organic farming*: removal of agrochemicals and inorganic fertilizers from farm inputs. Ecological farming practices are required to compensate for the loss of 'industrial' inputs.

In the context of the EC, all three types of farming system are already recognized under the CAP, Regulations 797/85, 1094/88 and 1272/88 respectively having been created to enable member states to introduce measures to subsidize the development of each system. Indeed diversified and low input-output farming systems are politically attractive: they reduce farm output or divert resources to products not in surplus, limit pressure on farm businesses to over-exploit their environmental resources, but do not require any fundamental change in the agroindustrial approach to farming. Measures to 'extensify' and 'convert' agriculture represent a politically acceptable approach to 'sustainable agriculture' in the EC. However, diversified and low input–output farming are more correctly described as 'alternative agricultures' (OECD, 1989) for they do not meet the challenge of significantly reducing the energy and agrochemical dependencies of modern farming (depletable resources), still require price (or income) support policies for their economic viability, and the costs of such policies continue to threaten the budgetary criterion (see iv above) of a 'sustainable' agriculture. In addition, modelling the impact of 'extensive' agriculture for farm business development, and estimating the consequences of non-targetted policies for the environment, have only just begun. Similarly, research on organic agriculture as an alternative path of farm business development has only recently gathered a momentum, but it can be argued that organic farming represents the only truly 'sustainable' type of agriculture in the long run.

Organic Agriculture as a Path of Farm Business Development: Experience in the United Kingdom

This section turns to the third 'ideal type' of sustainable agriculture – organic farming – to examine its recent development in the UK and so offer a perspective on its potential as an alternative path of farm business development. It is anticipated that individual farm businesses will continue to develop along a variety of paths (see above), so that organic farming is viewed as one type of business structure not as the totality of a national or regional farming system. The framework for the following observations is provided by the (i)–(v) criteria of 'sustainable agriculture' outlined above, while farm business information is taken from a postal questionnaire survey of all organic farms in the UK (accredited by the Soil Association and Organic Farmers and Growers) carried out during 1990/91. The survey yielded a 45% sample of useable replies (n=284) from the 633 farm businesses approached.

Environmental criteria

Organic agriculture, of all types of farming, comes closest to an ecologically based definition of 'sustainable agriculture'. Following the model of Hill and Ramsay (Troughton, 1991), organic farming is characterized by the following types of environmentally sensitive farming practices: a holistic approach to the farm system, the maintenance of ecosystem stability, 'stewardship' of the resource base, energy efficient technology, conservation of soil fertility through soil organic matter, improvement of food quality, crop rotations, mixed crop–livestock farming, absence of agrochemicals and inorganic fertilizers, use of animal and green manures, cyclical nutrient flows, and attention to animal welfare. Not surprisingly organic farming is viewed by its practitioners as an agricultural philosophy as well as a farm management system, and ethical considerations were certainly important for those pioneering commercial farmers who practised organic methods in North America in the 1960s and 1970s (Wolf, 1977). Moreover, the philosophical dimension is evident in the organic sector of UK agriculture today – in the farm survey, from the three main reasons given for undertaking organic farming, environmental and social responsibility (30% of reasons), food health (29%) and husbandry (22%) were placed ahead of economic (11%), personal (6%) and political reasons (2%). Moreover, it is the wider acceptance among consumers of 'healthy eating', and a concern with animal welfare, that has provided the market opportunity for organic producers in recent years.

Socioeconomic criteria

In Britain, as elsewhere in Western Europe, organic farming has had its advocates and practitioners for many decades, but only in the last 10 years has the farming system been taken up by significant numbers of commercial food producers. In the farm survey, for example, 16% of producers began their organic enterprises before 1970, but 61% have entered production since 1980 and 65% of farms have all their land in organic food production. Compared with all farm businesses in England and Wales, organic farms are drawn disproportionately from the small-size groups (with the exception of the 40–90 ha group), are more likely to be completely in owner-occupation, and tend to be farmed by younger people who have had a longer full-time education, but little farming experience. Of the organic enterprises 41% comprise both crop and livestock enterprises, with a further 37% specializing in horticultural (crop) production; over half (57%) of farm businesses marketed four or more products, thereby reflecting the diversified product base of their farming systems. On the trajectory of production, 40% of farm businesses had previously produced cereals and beef, whereas under organic farming these enterprises had fallen to a 29% representation, with horticulture and vegetable enterprises increasing to 38% (previously 21%). For a certain category of producers (defined above), therefore, organic farming offers a path of farm business development leading away from those sectors of production characterized by structural surpluses.

Turning now to a social issue in 'sustainable agriculture', organic farming has the capacity to generate new employment opportunities after several decades during which agriculture had consistently lost farmers and farm workers (Bowler, 1991). Research in a number of countries confirms the need for greater labour inputs under organic as compared with conventional farming practices (Lampkin, 1990, p. 512). Estimates from the farm survey (Table 17.1) indicate that casual hired work is most affected (an increase of 0.64 jobs per farm on average after the conversion from conventional farming), but full-time family work (0.41) and part-time hired work (0.47) also show significant increases. Unsurprisingly, the rate of employment generation is positively associated with farm size and horticultural/vegetable production; consequently, if the proportion of larger farms in organic agriculture increases, so the employment-generating capacity of the sector is likely to increase. In this way, organic farming can contribute to the stability of rural communities by stabilizing the employment base of agriculture. This is especially the case where organic producers are regionally clustered. In the United Kingdom, for example, Herefordshire/Worcestershire, Sussex, southwest Wales, Essex/Kent and Devon/Somerset/Dorset are the regions most likely to be favoured by the development of organic farming if the present distribution of such farms is a guide.

On the other hand, economic factors inhibit the development of organic farming as regards the costs of conversion and the net return from production. At present, organic farming products need to obtain an official accreditation (for example the Soil Association Symbol) so as to attract a premium on price. Such a premium is required for profitable production to offset the lower yields per hectare compared with conventional production. In the farm survey, 33% of farmers reported farm-gate price

Table 17.1. Employment change following the conversion to organic farming (284 sampled farms).

	Family			Hired			
	Full time	Part time	Casual	Full time	Part time	Casual	Total
Employment increase	137	104	64	103	139	228	775
Employment decrease	20	11	9	35	6	47	128
Net increase	117	93	55	68	133	181	647
Net increase (raised sample)	260	207	116	151	296	402	1438
Average increase in employment per farm	0.41	0.33	0.19	0.24	0.47	0.64	2.28

premiums of 10–20% while a further 25% reported premiums of 20–50%. But accreditation can take at least 2 years, during which time yields are lowered without the compensating premium on price. On the sampled farms, for example, 54% of occupiers had to obtain an off-farm income during the conversion period. It is expected that the EC will shortly announce a scheme of financial compensation for the conversion period and bring national schemes of support for organic farming under a standardized framework. But at present, especially for smaller farmers, the costs of conversion (income foregone) act as a resistance to the adoption of organic farming. On net returns per hectare of farmland, financial calculations for a variety of countries and farm types yield equivocal data (Lampkin, 1984 and Table 17.2). While all studies show a reduced yield per hectare for organic compared with conventional systems (an average of approximately 18% over crop and livestock production), the final financial outcome depends to a great extent on how the increased labour input is costed, and the margin of the premium on the farm-gate price. Nevertheless, most financial calculations demonstrate that organic farming does not consistently yield a lower profit margin compared with conventional systems, nor are the annual fluctuations in margins higher. Convincing conventional farmers of these realities, however, is a different matter; a high proportion of them are not yet prepared to take the risk of adopting a different system of farming. However, as the price of conventional products falls, and if the premium price for organic products can be maintained, so larger, conventional farmers are likely to convert a part of their farms to organic production on an experimental basis; they may then extend those organic operations if their results are favourable. But farmers facing the pressure of maintaining interest payments as a consequence of a high level of indebtedness seem least likely to take the risk of adopting organic farming practices.

Productionist criteria

A 'sustainable' agriculture must be capable of feeding the non-farm population. The lower crop and livestock yields generated by organic farming have attracted the criticism that a national farm sector based wholly on such farming would be unable to produce sufficient food for the domestic market; this view is expressed notwithstanding the surpluses of production found today in most sectors of agriculture. It is unrealistic to expect complete national farming systems to make a transition to organic farming, even in the medium term, given the dominance of agroindustrial systems. Indeed this chapter outlines the transition as feasible in the short to medium term for only certain categories and locations of farm businesses. Even so, the productivity of organic systems could be raised if similar resources on research and farm extension services were provided as for

Table 17.2. Comparison of the financial returns from organic and non-organic farming.

Location	Conventional	Organic	Unit
West Germany, 1989	1034	1142	DM/ha profit
Switzerland, 1979/81	115	93.9	SFr/person/day
England and Wales, 1988/89 (milk)[a]	1644	1299	£/ha gross margin
Switzerland, 1979/81 (milk)[a]	5826	5263	SFr/ha gross margin
United Kingdom*, 1989 (beef)[a]	898	630	£/ha gross margin
1989 (beef)[a]	336	88	
United States, 1974	314	324	$/ha net return
1975	346	333	
1976	333	336	
1977	278	289	
1978	384	333	

[a] No price premiums.
* Results from two separate surveys.
Source: After Lampkin (1990, pp. 502–33).

conventional farming. The underfunding of research on organic farming has been commented on widely (Wolf, 1977; Allen and van Dusen, 1989; Lampkin, 1990), but there is no reason why prospective developments in the biotechnology for crops and livestock could not be oriented to varieties and breeds less dependent on agroindustrial inputs.

Budgetary criteria

The prospect of reduced farm production is a positive attribute of organic farming, given the present system of linking state farm subsidies to the volume of output. Similarly, organic farming practices control many of the external costs of agroindustry by 'internalizing' those costs (Bateman and Lampkin, 1986). For example, cost savings could be made on present state-financed measures necessary to monitor and reduce pesticide and nitrate pollution, maintain 'traditional' farming practices, conserve landscape features and control the storage and disposal of farm slurry. However, as already indicated, there are increased costs to the farm budget of subsidizing the conversion of conventional farm businesses to organic production, while increased per unit output costs are passed on to consumers in the form of a premium on the price of organic food. As discussed in the next section, such price increases fall inequitably on the non-farm population even though there is a broad-based support for organically produced food.

Political criteria

It is commonly claimed that the balance of power in the formation of state farm policy is shifting away from producers and towards consumers/taxpayers (Bowler, 1985). At times it is difficult to discern such a shift, but a gradual reduction in farm support levels can now be detected, even in the EC. Surveys of consumer attitudes towards food and agriculture consistently reveal support for the reduction or banning of agrochemicals, an interest in buying organically grown produce and concern with the 'healthiness' of the food being consumed (Table 17.3). Support on these issues covers all age, socioeconomic and gender groups.

Translating these attitudes into actual consumer-buying behaviour, however, is more problematic, not least because of the price premium attached to organic food products. Numerous surveys have now been published which reveal that the retail price of organic products lies between 1.2 and 3.5 times above the price of conventionally produced food (Table 17.4). As a result, the purchasing of organic food is more prevalent in the upper compared with the lower socioeconomic groups, and among those who demonstrate other types of 'green' behaviour, for example through membership of environmental groups or recycling

Table 17.3. Attitudes to food health issues in the UK (samples of consumers).

'Artificial chemicals in agriculture should be banned/reduced'		71% agree
Interest in buying organically produced . . .	fresh vegetables	67%
. . .	fresh fruit	55%
. . .	fresh meat	50%
. . .	eggs	49%
. . .	milk	49%
'Food manufacturers should be more concerned about the dietary quality of their produce'		69% agree
'The countryside and the environment are being destroyed by today's farming methods'		60% agree

Abstracts from surveys by Mintel (1989) and Elm Farm Research Centre (1988).

household refuse. The main point, however, is that there is a latent support in society for organically oriented farming and the policies necessary to promote such an agriculture. To date, however, the development of organic farming has not been identified as a political issue; rather it is subject to benign neglect by most mainstream political parties in Western Europe.

Table 17.4. Price premiums on organic produce in supermarkets (UK).

Product	Safeway	Sainsbury	Gateway	Waitrose
Apples	× 2.5	× 3.5	–	× 2.6[a]
Pears	–	2.2	–	2.3
Carrots	–	2.2	1.7	2.1
Onions	–	2.9	1.9	1.7
Potatoes	1.4	1.7	1.6	1.6
Tomatoes	–	2.4	–	2.1
Cabbage	1.3	1.5	–	–
Mushrooms	–	1.5	–	1.3
Eggs	1.5	1.2	2.1	1.6
Bread	–	1.2	1.3	1.3
Chicken	1.3	1.4	–	1.5

[a] i.e. 2.6 times the non-organic price.
(Extract reprinted with permission from *Conservation Now* 1989, 1, 26–27.)

Organic Farming in the Food Supply System

The literature on 'sustainable' agriculture, including organic farming, tends to concentrate on the production sector; relatively little attention is given to the place of organic producers within the whole food supply system. But based on the survey of organic farms in the UK, it is clear that an alternative food supply system is being created for organic produce;

alternative, that is, to the system created for the 'industrial' model of farm production. There appear to be two reasons for this development: first, the sector dealing with inputs to farming has not responded to the needs of organic farmers because of the lack of technical knowledge and the comparatively small size of the market. Consequently a range of relatively small firms has been able to set up to provide inputs such as: seeds, machinery (weeders, cultivators, etc.), crop covering materials, composts, animal feed and organic fertilizers. Only recently have training courses on organic farming been provided by a number of existing farmers and agricultural colleges, while information on organic farming from specialist advisers has been made available through the Agricultural Development and Advisory Service (ADAS) only in the last 2 years. A second constraint in the conventional food supply system has been provided by the large supermarket chains who have imposed on organic produce the same demands for quantity, quality and continuity of supply, as well as grading and packaging, as for conventionally produced food (Daw *et al.*, 1991). Supermarkets have tended to deal directly with larger growers and farmer cooperatives. In addition, the pricing policy of the supermarkets (see price premiums in Table 17.4), the amount of shelf space devoted to organic produce, and the display of such food, has not encouraged consumers to buy non-conventional foods. By 1990, organic food sales were still valued at less than £50 million a year in the UK (less than 1% of the retail food market). After negative experiences with supermarkets, many farmers have turned to alternative outlets such as farm shops (10% of the farm sample), farm gate sales (23%), market stalls (2%) and independent fruit/vegetable retail and 'health' food shops (20%). But there is considerable scope for domestic (UK) producers to supply supermarket outlets: 60% of organic food is still imported from Germany, France and The Netherlands by companies such as Geest.

Problems over obtaining supplies from producers within the UK have also dissuaded many existing wholesale firms from becoming involved in the marketing of organic produce. Consequently new wholesalers have been established to specialize in the organic sector, whereas farmers themselves have grouped into regional cooperatives to market their produce (23% of the sampled farms – for example, Organic Farmers and Growers (Suffolk), Green Growers Co-operative (Herefordshire) and Organic Growers West Wales). These features of the food supply system have developed mainly in those regions with large numbers of organic farms; such developments tend to intensify the uneven spatial development of this type of 'sustainable' agriculture. The general point from these observations is that an alternative food supply system, or significant change to the existing system, is needed to support the development of a 'sustainable' agriculture in the longer term.

Conclusions

Most informed observers recognize that the agricultural sector in developed market economies has reached a crisis in its evolution. In summary form, the crisis contains the twin dimensions of 'production' and 'environment'. In North America, the reaction among analysts and policy advisers has been to develop the concept of a 'sustainable' agriculture, recognizing that ecologically based farming systems offer the best option despite the confrontation with contemporary agroindustrial systems. In Western Europe, by comparison, the reaction has been to cast the debate in the context of 'alternative agricultures', with attention given to reducing harmful agro-inputs, diversifying the use of farmland and subsidizing the development of 'environmentally friendly' farming practices. There is less concern about a fundamental review of agroindustry, with more attention given to politically acceptable palliatives for the present crisis. Indeed merely reducing the intensity (volume) of production in the whole farm sector is viewed as a main aim of agricultural policy. Ecologically based farming, for example organic farming, has not been the subject of coherent state farm policies; rather this type of farming has developed in the main independently of state intervention.

As in North America, organic farming in Western Europe has been developed as an economically viable path of farm business development by only a minority of producers (22 000 ha registered area in the UK by 1990). The lower yields of the system are compensated in part by cost savings on agro-inputs but more especially by price premiums at the farm gate. In addition, the system of organic production is supported by a minority of consumers in the upper socioeconomic groups who have the income to be able to convert their 'green' ethics into actual food-purchasing behaviour. For organic production to develop, the price premium on organic food will have to be reduced thereby converting latent into actual demand among families on lower incomes.

In this respect, therefore, attention has been turned to the food supply system as a whole, and the role of organic (sustainable) agriculture within it. In the UK at least, organic farming has had to develop its own food chain parallel to that which has been developed to provide the mechanical equipment, seeds, crop and animal protection materials, composts, organic fertilizers and technical information required by organic farmers and growers. At the same time, organic producers have had to form themselves into marketing cooperatives in order to provide the continuity of quality and volume of supply required by wholesalers and retailers. Lastly, retail outlets alternative to supermarket chains have been developed; they enable producers to resist the imposition of the 'industrial model' of production by supermarket chains as regards their demands on price, quality, uniformity and packaging of farm produce. As a result, an alternative 'sustainable' food supply system is slowly emerging despite, rather than because of, state agricultural policies.

References

Allen, P. and van Dusen, D. (1989) Sustainable agriculture: choosing the future. In: Allen, P. and van Dusen, D. (eds), *Global Perspectives on Agroecology and Sustainable Agricultural Systems*. University of California, Santa Cruz, pp. 1–13.

Bateman, D. and Lampkin, N. (1986) Economic implications of a shift to organic agriculture in Britain. *Agricultural Administration* 22, 89–104.

Bijman, J., van den Doel, K. and Junne, G. (1986) *The International Dimension of Biotechnology in Agriculture*. European Foundation for the Improvement of Living and Working Conditions, Dublin.

Bird, E.R. (1989) Why modern agriculture is environmentally unsustainable. In: Allen, P. and van Dusen, D. (eds), *Global Perspectives on Agroecology and Sustainable Agricultural Systems*. University of California, Santa Cruz, pp. 31–7.

Bowler, I.R. (1985) Some consequences of the industrialization of agriculture in the European Community. In: Healey, M.J. and Ilbery, B.W. (eds), *The Industrialization of the Countryside*. GeoBooks, Norwich, pp. 75–98.

Bowler, I.R. (1991) The agricultural pattern. In: Johnston, R.J. and Gardiner, V. (eds), *The Changing Geography of the United Kingdom*. Routledge, London, pp. 83–114.

Bowler, I.R. and Ilbery B.W. (1992) *Farm Diversification in England and Wales*, Occasional Paper 21, Department of Geography, University of Leicester.

Brklacich, M., Bryant, C.R. and Smith, B. (1990) Review and appraisal of the concept of sustainable food production systems. *Environmental Management* 15, 1–14.

Commission of the European Communities (1989) *The Impact of Biotechnology on Agriculture in the European Community to the Year 2005*. The Commission, Brussels.

Cox, G., Lowe, P. and Winter, M. (1990) The political management of the dairy sector in England and Wales. In: Marsden, T. and Little, J. (eds), *Political, Social and Economic Perspectives on the International Food System*. Avebury, Aldershot, pp. 82–114.

Daw, M., Slee, B. and Wynen, E. (1991) *Organic Agriculture: A Review of the Marketing and Economics of Production with Particular Reference to Scotland*. Economic Report 32, Scottish Agricultural College, Aberdeen.

Goodman, D. and Redclift, M. (eds) (1989) *The International Farm Crisis*. Macmillan, London.

Goodman, D., Sorj, B. and Wilkinson, J. (1987) *From Farming to Biotechnology: A Theory of Agro-industrial Development*. Basil Blackwell, London.

Hill, S.B. (1985) Redesigning the food system for sustainability. *Alternatives* 12, 32–6.

Ilbery, B.W. (1990) Adoption of the arable set-aside scheme in England. *Geography* 76, 69–73.
Jackson, W. (1989) Ecosystem agriculture: the marriage of ecology and agriculture. In: Allen, P. and van Dusen, D. (eds), *Global Perspectives on Agroecology and Sustainable Agricultural Systems*. University of California, Santa Cruz, pp. 15–19.
Jones, A. (1991) The impact of the EC's set-aside programme. *Land Use Policy* 8, 108–24.
Keeney, D.R. (1989) Toward a sustainable agriculture: need for clarification of concepts and terminology. *Journal of Alternative Agriculture* 4, 101–5.
Lampkin, N. (1984) Studies of biological farming systems in Western Europe and North America – a literature review. In: Vogtmann, H., Boehncke, E. and Fricke, I. (eds), *The Importance of Biological Agriculture in a World of Diminishing Resources*. Happ Burkhard, Witzenhausen, pp. 123–49.
Lampkin, N. (1990) *Organic Farming*. Farming Press, Ipswich.
Merril, R. (ed.) (1976) *Radical Agriculture*. Harper and Row, New York.
Organisation for Economic Co-operation and Development (1989) *Agricultural and Environmental Policies: Opportunities for Integration*. OECD, Paris.
Pierce, J.T. (1990) *The Food Resource*. Longman, London, pp. 297–313.
Potter, C. (1986) The environmental effects of CAP reform. *Countryside Planning Yearbook* 7, 76–88.
Robinson, G.M. (1991) EC agricultural policy and the environment: land use implications in the UK. *Land Use Policy* 8, 95–107.
Troughton, M.J. (1991) Ecological assessment of modern agriculture. In: van Dort, G., van den Berg, L., Groenendijk, J. and Kempers, A. (eds), *Limits to Rural Land Use*. Pudoc, Wageningen, pp. 141–202.
Whatmore, S., Munton, R., Little, J. and Marsden, T. (1987) Towards a typology of farm businesses in contemporary British agriculture. *Sociologia Ruralis* 37, 21–37.
Wolf, R. (1977) *Organic Farming: Yesterday's and Tomorrow's Agriculture*. Rodale Press, Emmaus.

18

Alternative Agriculture and Conventional Paradigms in US Agriculture

Janel Curry-Roper

Societal norms and values of the 18th and 19th centuries have been influential in the development of the conventional paradigm in United States agriculture. This paradigm includes such elements as the centralization of marketing and processing, individual self-sufficiency, primacy of private property, and the reducibility of the farm into individual components through the specialization and standardization of production (Table 18.1). This conventional paradigm, also described as the 'industrial model' of agriculture (see Chapter 2), assumes the present scientific paradigm as applied to agricultural problems.

These values and norms are grounded in the Enlightenment philosophy of Locke and Smith with its emphasis on rationalism, reductionism and the individual. Locke maintained that society is a collection of individual beings and that the foundation of society is the will of the individual (Zylstra, 1981, p. 14). Locke tied that individual will to the acquisition of private property. In his Second Treatise on Government, he taught that humans have property in their own persons, in the labour of their hands, and that the earth's resources can only fulfil their needs if it belongs to them, apart from anyone else (Zylstra, 1981, p. 16). In other words, a person was always ultimately separate from a whole and could only find expression and fulfilment in private ownership.

Similarly reductionistic was Smith's model of economic development: the economy emerges from the free actions of autonomous, self-sufficient individuals. Value (i.e. the value of goods), is not in their use, but only in (their value in) exchange between these autonomous individuals. Value thus must be tied to the market (Woods, 1985, p. 7).

Thomas Jefferson became the originator of and advocate for the United States land policy which reflected Enlightenment ideals. He advocated the building of a nation of self-sufficient, independent farmers in tune with the earth that supported them (Petulla, 1977, p. 106). This was

Table 18.1. Agricultural paradigms.

Conventional agriculture	Alternative agriculture
Centralization	*Decentralization*
National/international production, processing, marketing	Local/regional production, processing, marketing
Concentration of resources	Dispersed resources
Fewer farms	More farmers
Individualism and Competition	*Community*
Self-interest	Increased cooperation
Reduced labour	Meaningful labour
Farming as business	Farming as way of life
External costs ignored	All costs considered
Material success	Non-material values
Reducibility of farm into individual components	*Farm as system*
	Imitation of natural ecosystems
Specialization	Diversification
Standardized production	Localized production systems
Fine Tuning of present scientific paradigm	*Radical reconception of science*

Source: Adapted from Beus and Dunlap (1990, pp. 598–99).

accomplished with an emphasis on economic rationalization as expressed through the United States National Land Survey, whose grid pattern was perfectly suited for quick disposal of public lands to individuals. This way of life, represented by the separate farmstead, was considered the norm, and developers of the Ordinance of 1785 took it as its model. It was confirmed by the Pre-emption Act of 1841 which required one to live on the land he/she claimed. Land became another commodity that was valued for its largely economic possibilities. Historical constraints on individual materialistic gain were minimized and American law began to assume the superiority of individual private property over every other social good (Petulla, 1977, p. 88).

This cultural ideal, with the emphasis on the individual and land outside of community, permeated society. It is within the framework of this ideology that US policy has been formed and constrained in the past. Federal farm policy assumes that the farm economy emerges from the free actions of autonomous, self-sufficient individuals and thus targets the individual. The primacy of private property rights has also been assumed, so policy has been based on voluntary participation in farm programmes encouraged by monetary incentives.

Sustainable Agriculture within the Conventional Paradigm

One example of a problem facing farming in the United States, in which the conventional agricultural paradigm frames the issue, is the criticism of farming as the source of non-point surface and groundwater pollution from pesticides, nitrates and sediment (National Research Council, 1989, p. 3). Farmers, researchers and policy-makers have been faced with the question of how to affect change and reduce this environmental pollution. What have been the assumptions and the responses to this concern within the farm community that remains tied to the conventional ideology of farming?

Locke's teaching promised individual freedom, opportunity to compete for unlimited individual material wealth, and the limitation of government to interfere with individual initiative. It should come as no surprise, therefore, that individual voluntary compliance and incentives have been emphasized along with technical research in economically viable alternative methods of farming. These alternatives, called 'sustainable' or 'low-input' agriculture, emphasize better management and reduced costs. Research in alternative agricultural systems has continued to be dominated by the traditional Enlightenment scientific worldview of rationalism and reductionism, while critics question whether conventional and alternative agricultural systems can be described and analysed using the same scientific theories, methods and concepts as past agronomic research. Keller (1985), who critiques science from a feminist perspective, points to some of the underlying assumptions of science and their limitations. Her critique seems to be particularly applicable to the problems facing sustainable agriculture.

At present, the research in sustainable agriculture remains reductionist and causational. The variables to be studied are reduced in the attempt to carry out a controlled experiment that is quite abstracted from the reality of a farm system, but can be scientifically validated within the present scientific paradigm. On-farm experiments are shied away from because of their lack of scientific validation.

Keller presents an alternative approach to science that seems to better fit the farm situation. She proposes order as a category organizing scientific inquiry, as opposed to the more restrictive emphasis on scientific laws that imply external constraints (1985, p. 132). She sees an emphasis on order causing a shift from the more static, causational models to interactive models of systems – a scientific approach no less rigorous but presupposing the modesty and open attentiveness that allow one to 'listen to the material' (1985, p. 134).

Keller sees this search for order as assuming an *a priori* complexity that exceeds our human imagination, an assumption that does not allow factors to be considered to be limited, nor for questions like the nature of the systems and our relationship to them to be discounted (1985,

p. 147). It is an approach that is not predictive or causational, but attempts to gain an understanding that includes human connections with the world. If a paradigm like Keller's was adopted, less emphasis would be placed on reducing elements to their basic components and more on describing the order of an entire system.

Scientists also have the problem of the non-scientific elements of value and commitment in agricultural research. Social and natural scientists involved in sustainable agricultural research find it difficult to deal with the question 'where do we go from here?' because of the assumption of objectivity. An example of this within sustainable agriculture is the inability to deal with the issue of land tenure because of its value-laden nature, though it may impact directly on land use. McGuire, a sociologist, describes this problem as one where science has attempted to eliminate the element of commitment from human speech. He argues that 'speech by its nature reflects commitment, residing in the realm of "ought", such that social scientific attempts at purification have been necessarily unsuccessful' (unpublished). McGuire claims that substantive discourse on issues ended with the assumption that somehow data overcome differences in opinion and values. Keller points to the same problem in science:

> The ideology of modern science, along with its undeniable success, carries within it its own form of projection: the projection of disinterest, of autonomy, of alienation . . . the dream of a completely objective science . . . The objectivist illusion reflects back an image of self as autonomous and objectified: an image of individuals unto themselves, severed from the outside world of other objects and simultaneously from their own subjectivity. It is the investment in impersonality, the claim to have escaped the influence of desires, wishes, and beliefs – perhaps even more than the sense of actual accomplishment – that constitutes the special arrogance, even bravura, of modern man, and at the same time reveals his peculiar subjectivity.
>
> (Keller, 1985, p. 70)

By not confronting the question of where to go from here, scientists have accepted a value system that is assumed – the conventional one – and have accepted it as if it were objective.

Within sustainable agriculture, education and outreach programmes have had the same basic assumptions as scientific research. They have conceptualized the problems of agriculture on the individual level and thus targeted individuals, assuming the individual as the main building block of society. The individual is seen to respond only to immediate economic interests with no constraints of a larger community.

The final result to date has been the institutionalization of the positive environmental image of sustainable agriculture within the Land Grant Universities, the United States Department of Agriculture and, more recently, agrochemical firms, without any fundamental change in the basic conventional ideology underlying the farm enterprise (Beus and Dunlap, 1990, p. 612). The sustainable agricultural movement has continued to reduce the natural and human aspects of sustainability to their individual components.

While agriculturalists come from the perspective of voluntary compliance and incentives, environmentalists, who are now seeing agriculture as just another business that must meet certain standards, emphasize regulation (Zinn and Blodgett, 1989, p. 185). In actuality both these attitudes are a reflection of the same conventional ideology; the ideology of individualism, as we have seen, has emphasized private ownership, speculation, and resource as commodity outside of community. These attitudes have limited the solutions to farm problems in the United States to two options: either private individuals acting in a free market responding only to economic incentives have been seen as the solution to the problem; or the federal government has been called on to intervene, targeting individuals in its intervention. One is but the mirror image of the other; one necessitates the other. When the effects of the actions of such individual interests have caused third-party problems (water pollution, erosion, higher prices) we have tended to call for federal intervention to solve the problem through regulation. Where the state is perceived as only the protector of individual interests, it becomes inevitable that centralized state functions are required to balance the forces of individual interests. It appears that a particular model of the state has been presupposed – one derived from such thinkers as Locke and the neoclassical economists, such as Von Mises. The function of government is to maintain the social system that allows the individual the freedom to pursue his or her own best interest, usually expressed in private property. 'Good' actions by the state are those that allow the individual to attain the ends he or she wishes to attain.

Environmentalism has attempted to transcend these individualistic tendencies, accepted as given, by appealing to a more universal commitment to the earth. Still within the individualistic mode, Lasch (1991, p. 36) calls this approach ineffective because it stretches our capacity for loyalty too thin. Such commitments need to be attached to specific people and places, not to an abstract ideal.

Society as Individuals Versus Society as Social Groups

Sociologists and scholars who work on common resource management offer insights into community-wide patterns of values affecting farming –

the level at which commitments operate. For both sets of scholars, solutions to resource and agricultural problems lie in the renewal of community and of institutions that make it possible and vital (Bellah, 1991). Their research has found that community-wide social patterns that affect the farm system do exist (Parsons and Waples, 1945; Flora and Stitz, 1985). They exist in spite of the strong ideology of atomism: the belief that the farm economy emerges from the free actions of autonomous, self-sufficient individuals, and in spite of a legal and ideological climate that reinforces these beliefs. The community-wide social patterns are especially evident among the ethnically and religiously homogeneous communities that dot the rural agricultural regions of the Midwest. They influence capitalization of the farm enterprise, the extent to which a farm is commercialized and farmers' risk-reduction strategies (Flora and Stitz, 1985; Salamon and Davis-Brown, 1986). More importantly, these value systems have led to community-wide patterns of strategies for intergenerational land transfers (Salamon, 1980, 1984; Rogers and Salamon, 1983; Salamon *et al.*, 1986), the effectiveness of which is the sustainability of the farm unit. Inasmuch as land use is tied to ownership, the continued re-creation of the farm enterprise may be related here as well.

The idea of isolation of community-wide patterns is well illustrated by a recently completed study on 10.25 km^2 (4 square miles) in south central Iowa. The study focused on all land transfers within this ethnically Dutch area from settlement to the present (Curry-Roper and Bowles, 1991). After initial settlement of the study area in the early 1850s, community-wide eras of generational transfers appeared to be the primary force behind the land-transfer cycle. These decades of transfers occurred approximately every 30 years, interrupted only by the First World War inflation and the following Depression. Peak time periods in farm transfers involved generational transitions, whereas non-family transfers dominated in times of low activity.

The Dutch studied in south-central Iowa had unique inheritance characteristics of both Yankee and Yeoman farmers – Midwest community types as defined by Salamon (1985, p. 326). The Yankee farmer viewed farming as a business and had no obsession with transferring the land to the next generation as a farm unit. The Dutch farm owners in the study area shared this trait, perhaps due to the fact that many who settled in the area were not farmers in the Netherlands. Very few inheritances occurred before the next generation were in their fifties. Farm families were found to follow the practice of 'partible' inheritance, where land was divided among heirs (Salamon, 1980, p. 291).

Within the seven south-central Iowa families studied, there were few pre-death transfers. Evidently the offspring of these Dutch settlers had to enter farming by purchasing their own land, or farm as tenants on their parents' farms. Many offspring chose other professions altogether since the Dutch emphasized education, upward mobility and movement to non-farm occupations.

The Dutch settlers in the area also exhibited traits of the German Yeoman Farmers, as described by Salamon (1980): strong community attachments, with little out-migration, endogamy and ethnically homogeneous communities. Evidently the Dutch combined these two value systems by developing local industry and business, thus emphasizing both upward mobility away from the farm and maintenance of a generally homogeneous community – the town of Pella (Doyle, 1985; see examples in Bjorklund, 1964; Kirk and Kirk, 1974). These characteristics have made the area less dependent on farming today, and the entrepreneurial spirit has led to the development of several family-owned manufacturing companies that do business in the international market.

Persistence of land holding, subject to much rhetoric, may also be very culturally determined. Persistence was low within the study area: less than 1% of the land has remained within the same family from 1850 to the present (8 ha) and this passed from uncle to nephew rather than directly. An analysis of 50-year time blocks, each spanning two generations in the family life-cycles, showed surprising similarity in lack of persistence. Only about 20% of the land stayed in one family for each of the two 50-year blocks of time (1850–1899 and 1900–1949). Most land passed among two to three families within each 50-year time period. This pattern is another example of non-yeoman behaviour that may have been tied to upward mobility (Flora and Stitz, 1985, p. 349).

Alternative Legal Frameworks for the Agricultural Community

The previously discussed studies show that community-wide value patterns exist and affect farming and the local entrepreneurial climate. Indeed scholars interested in the management of common resources have begun to look more closely at legal frameworks that enable or disable communities in the management of their resources and thus their efforts at environmental control. The definition of rights in property is a crucial area of research when dealing with issues of rural restructuring, as Whatmore *et al.* (1990) have illustrated. Though we do not usually think of farm enterprises having anything to do with common property regimes, the concepts now being developed in this area have much to say relative to the issues facing farming in the United States.

Berkes and Feeny (1990, p. 48) have pointed out that the common property paradigm attempts to address philosophies that underlie resource management. It examines not just the nature of the resource but the institutional conditions and cultural values that lead to sustainability.

Whereas Hardin, in his 'Tragedy of the Commons', assumed individual interest that was not moderated by the surrounding community, and thus the necessity for force from the highest level, the commons paradigm assumes the possibility of cooperation at the community level. This possibility is further substantiated by the evidence that community-level patterns of behaviour and values contine to exist in the United States. Whereas Adam Smith assumed order coming naturally from individuals pursuing their own self-interest, one can alternatively assume the possibility of order arising from conscious collective action towards a common good (Berkes and Feeny, 1990, p. 50). This collective action must be at the level at which commitments are made.

As with common property resources, the farming community, in its social groups, is cut off from the legal aspects of property rights; the rights of the individual and the state by contrast have been upheld. In fact, attempts to continue a communal form of life by immigrant groups has been undermined by legal forces. In this way, informal community social controls have been undermined by the lack of legal recognition. The concept of legal pluralism, or the co-existence of multiple legal systems in a society, might be applied to this situation in the way that indigenous law and state law co-exist (Fortmann, 1990, p. 197).

New Paradigms within the Agricultural Community

Somewhat outside the established order, a variety of movements have emerged that are similar to one another in the agricultural paradigm they support – decentralization, emphasis on community, and radical reconceptualizations of science (Table 18.1). Among agricultural researchers, Jackson has become a major spokesman. Proposing a science more akin to Keller's, Jackson (1987) sees the need to be more mindful of the original material of nature rather than the work of the scientist:

> What if we researched and taught as though we believed that the wisdom of nature is more important, in the long run, than the cleverness of science? . . . What if we acknowledged straight out that there is more to be discovered than invented? Of course we must have both discovery and invention, but what if we changed the emphasis?
>
> (Jackson, 1987, p. 10)

At his Land Institute in Kansas, Jackson is attempting to develop a mixture of perennial grain crops that simulates the prairie, and the science that is emerging emphasizes systems. This approach is different from traditional science in that a systemic whole cannot be predicted from its individual parts; causation within it moves downward, with cause and effect often being indirect and non-linear; and emphasis is placed on relationships as primary factors within the system, rather than just individual objects (Callicott, 1990, pp. 42–44). The bioregional movement has a similar worldview. It is a popular, loosely defined movement that attempts to divide the country into natural regions, and build a self-sustaining way of life based on the natural and cultural characteristics unique to a region (Zuckerman, 1987).

A similar feature of these attempts to redefine our way of understanding agricultural systems is that they include a critique of the individualistic emphasis of American society. Jackson (1987, p. 68) connects the social ideology that assumes the individual prior to society with the same assumptions in science. Although the ability of those that hold the conventional paradigm (United States Department of Agriculture, Land Grant Universities, etc.) to adopt the politically attractive elements of sustainable agriculture, while ignoring the radicalness of its core, should not be underestimated, what may be taking place is a shift in paradigm, as described by Kuhn (1970). The conventional networks of commitments – conceptual, theoretical and methodological – are being challenged. The challenge comes in response to a crisis that results in the articulation of alternative paradigms and discussion of philosophical assumptions underlying present approaches. Kuhn (1970, p. 91) describes these as the symptoms of transition.

What makes this challenge unique is that it questions not only the scientific paradigm within which agricultural research has been taking place, but also the assumptions on which American society is based, both of which go back to the Enlightenment. The comprehensiveness of the challenge makes the paradigm more radical but less likely to be fully accepted. And while the challenge within the natural science aspect of agriculture is gaining depth and credibility, the alternative societal paradigm has yet to truly get beyond rhetoric and articulate alternative ways of formulating a farm policy and conceptualizing farm problems. Those that support alternative societal paradigms in agriculture could gain much from common property scholars, but may have limited impact within the context of the growing move toward international free trade and competition. Whether an alternative scientific paradigm can transform our culture, like the scientific paradigm of the Enlightenment, which shaped the farm enterprise in all its aspects, has yet to be seen.

References

Bellah, R.N. (1991) The rise of market totalitarianism. *The New Oxford Review* 58, 8–15.

Berkes, F. and Feeny, D. (1990) Paradigms lost: changing views on the use of common property resources. *Alternatives* 17, 48–55.

Beus, C.E. and Dunlap, R.E. (1990) Conventional versus alternative agriculture: the paradigmatic roots of the debate. *Rural Sociology* 55, 590–616.

Bjorklund, E.M. (1964) Ideology and culture exemplified in Southwestern Michigan. *Annals of the Association of American Geographers* 54, 227–41.

Callicott, J.B. (1990) The metaphysical transition in farming: from the Newtonian-mechanical to the Eltonian-ecological. *Journal of Agricultural Ethics* 3, 36–49.

Curry-Roper, J.M. and Bowles, J. (1991) Local factors in land tenure change patterns. *Geographical Review* 81, 443–56.

Doyle, R.L. (1985) Wealth mobility in Pella, Iowa, 1947–1925. In: Svierenga, R.P. (ed), *The Dutch in America*. Rutgers University Press, New Brunswick, pp. 156–71.

Flora, J.L. and Stitz, J.M. (1985) Ethnicity, persistence, and capitalization of agriculture in the Great Plains during the settlement period: wheat production and risk avoidance. *Rural Sociology* 50, 341–60.

Fortmann, L. (1990) Locality and custom: non-aboriginal claims to customary usufructuary rights as a source of rural protest. *Journal of Rural Studies* 6, 195–208.

Jackson, W. (1987) *Altars of Unhewn Stone*. North Point Press, San Francisco.

Keller, E.F. (1985) *Reflections on Gender and Science*. Yale University Press, New Haven.

Kirk, G.W. and Kirk, C.T. (1974) Migration, mobility and the transformation of the occupational structure in an immigrant community: Holland, Michigan, 1850–1880. *Journal of Social History* 7, 142–64.

Kuhn, T. (1970) *The Structure of Scientific Revolutions*. University of Chicago Press, Chicago.

Lasch, C. (1991) *The True and Only Heaven*. W.W. Norton, New York.

National Research Council (1989) *Alternative Agriculture*. National Academy Press, Washington DC.

Parsons, K.H. and Waples, E.O. (1945) *Keeping the Farm in the Family*. Wisconsin Agricultural Experiment Station Bulletin 157, Madison.

Petulla, J.M. (1977) *American Environmental History: The Exploitation and Conservation of Natural Resources*. Boyd and Fraser, San Francisco.

Rogers, S.C. and Salamon, S. (1983) Inheritance and social organization among family farmers. *American Ethnologist* 10, 529–50.

Salamon, S. (1980) Ethnic differences in farm family land transfers. *Rural Sociology* 45, 290–308.

Salamon, S. (1984) Ethnic origin as explanation for social land ownership patterns. *Research in Rural Sociology and Development* 1, 161–86.

Salamon, S. (1985) Ethnic communities and the structure of agriculture. *Rural Sociology* 50, 323–40.

Salamon, S. and Davis-Brown, K. (1986) Middle-range farmers persisting through the agricultural crisis. *Rural Sociology* 51, 503–12.

Salamon, S., Gengenbacher, K.M. and Penas, D.J. (1986) Family factors affecting the intergenerational succession to farming. *Human Organization* 45, 24–33.

Whatmore, S., Munton, R. and Marsden, T. (1990) The rural restructuring

process: emerging divisions of agricultural property rights. *Regional Studies* 24, 235–45.
Woods, D.J. (1985) 'Nature' in Adam Smith's *Wealth of Nations*. Paper series, Institute for Christian Studies, Toronto.
Zinn, J.A. and Blodgett, J.E. (1989) Agriculture versus the environment: communicating perspectives. *Journal of Soil and Water Conservation* 44, 184–7.
Zuckerman, S. (1987) Living There. *Sierra* 72, 61–7.
Zylstra, B. (1981) The individual gospel. In: Harper, W.A. and Malloch, R.R. (eds), *Where Are We Now?* University Press of America, Washington DC, pp. 11–19.

CONCLUSION

It is probably inherent to the human condition that every generation considers itself to be living through a period of significant and rapid social and economic change. But only with the benefit of hindsight allowed to later observers of society can a true appreciation be gained of earlier events. From this perspective, the significance attached to the changes (transition) in agriculture reported in this book may prove to be exaggerated and to have rather less moment than claimed. On the other hand, when taken together, the evidence presented in the various chapters is persuasive of the view that we are living through a turning point in the development of the agricultural sector under advanced capitalism, at least so far as its place in the food chain and its interaction with the environment is concerned.

The origins of this book – in the papers of a conference – make it inevitable that some themes are repeated as various authors develop the logic of their arguments. However, insofar as the contributors have been drawn from three different cultural backgrounds (Canada, the United States and the United Kingdom), each with its own value system and assumptions, any repetition is itself of interest. It helps us to identify those commonalities that are found in the agricultural sectors of all developed market economies under the contemporary restructuring of agriculture. In this respect, four main conclusions can be drawn from the 18 chapters of this book.

First, it is clear that national agricultures have been drawn irreversibly into a global food supply system. It can be argued that this is not a new development: for so long as there has been international trade in food there has been a global supply system. But a qualitatively and quantitatively new system has developed: it is based on the industrialization of agriculture (Chapters 1–3) and the absorption of farms into the food chain as suppliers of raw materials to a food-processing sector organized within the context of transnational corporations. These corporations not only seek their raw materials on an internationally competitive market, but their products are similarly marketed competitively at a global scale. Farms and their agricultural regions are having to reorganize to compete among themselves not just in a national context: they are also having to compete internationally. This trend will be reinforced should ongoing trade negotiations under GATT result in the liberalizing of world trade in farm products. In this competitive environment, a relatively few but large production units (farms) are able to supply the required commodities, thereby rendering redundant the majority of smaller family farms. All societies with developed market economies are now faced by the dual problems of redundant farm families and the land which they occupy.

Second, no alternative use for the (surplus) factors of production on farms (land, labour and capital) is without its problems, including placing the land under trees (Chapters 4–7). Research reported in this book confirms that farm diversification offers only regional niche markets, whether producing farm commodities or non-farm products and services, whereas only a conversion to woodland/forest appears capable of absorbing the surplus farmland. But such a development would be unable to support present numbers of farm families, at least in the short term.

Third, as a challenge to conventional modes of thought, the concept of reorganizing agriculture as a 'sustainable' system of land use is increasingly accepted. Adoption of a 'sustainable' agriculture would have the capacity to keep farm families on the land and counteract the damaging consequences of contemporary farming practices. Indeed acceptance of the need for a more sustainable agriculture has been shaped, in large part, by the mounting evidence on the external costs of contemporary farming practices, costs that societies in many countries seem no longer willing to bear (Chapters 8–10). But 'sustainable' agriculture is itself problematic, not least because there is no internationally agreed definition, and the implications of applying such a system at the level of the individual farm business have yet to be thoroughly researched (Chapters 16–18). Moreover, the widespread adoption of 'sustainable' agriculture would require accceptance of a paradigm of agriculture that is the antithesis of the presently dominant conventional paradigm. There are many vested interests to be overcome in moving away from the 'industrial' model of agricultural development.

Fourth, state farm policies have the capacity to exert a major influence over such issues as the use to which surplus farmland is to be put, the fate of the family farm sector, the ways in which environmentally friendly farming practices are introduced, and the extent to which 'sustainable' agriculture is developed within the farm sector. Evidence presented in this book (Chapters 11–15) suggests that governments have not as yet formulated clearly defined programmes to react to the contemporary restructuring of agriculture. Individual policy measures appear more as a reaction to particular pressures rather than the product of a carefully designed strategy for meeting the problems – social, economic and environmental – of a restructured agriculture.

Finally, this book has highlighted two dimensions that have been obscured in other recently published books on changes in contemporary agriculture. One is the reintroduction of 'nature' into the debate on the restructuring of agriculture. Theorists, particularly within political economy, have tended to emphasize social and economic factors in their analyses, reducing 'nature' to a contextual role. This book has shown how 'nature' has placed limits around and opportunities before those who are employed in agriculture and has thereby shaped the contemporary restructuring of the sector. A case can be made for the need to integrate 'nature' within theorizations of the contemporary transition of the agricultural system.

The second highlighted dimension is the spatially varying impact of change in agriculture and the relationship with the environment. Whereas the geographical (or regional) context may lack the interest of theorists, for the farm families in agriculture the regional impact of restructuring is a vital concern. As ever, policy makers also tend to avoid considering the varying regional needs of agriculture under the restructuring process, but the chapters in this book make the case for greater attention to the regional context of change in the agricultural system.

Index